Edward John Hamilton

The Modalist or the Laws of Rational Conviction

A Text Book in Formal or General Logic

Edward John Hamilton

The Modalist or the Laws of Rational Conviction
A Text Book in Formal or General Logic

ISBN/EAN: 9783337233006

Printed in Europe, USA, Canada, Australia, Japan

Cover: Foto ©berggeist007 / pixelio.de

More available books at **www.hansebooks.com**

THE MODALIST

OR

THE LAWS OF RATIONAL CONVICTION

A TEXT-BOOK

IN

FORMAL OR GENERAL LOGIC

BY

EDWARD JOHN HAMILTON, D.D.

Albert Barnes Professor of Intellectual Philosophy in Hamilton College, N.Y.

BOSTON, U.S.A.
PUBLISHED BY GINN & CO.
1891

TYPOGRAPHY BY J. S. CUSHING & CO., BOSTON, U.S.A.

PRESSWORK BY GINN & CO., BOSTON, U.S.A.

PREFACE.

This text-book was written under the conviction that the most useful instruction is that which is enforced by the most thorough explanations. It is an attempt to connect the formulas of logic with principles, the ultimate character of which will become evident to the faithful student. Besides, the author had an ambition to add something to the science by giving permanent form to views which have been held and taught for years.

Logical doctrine and praxis do not now have that place in education which they once had, when the university curriculum was chiefly occupied with the literature and the philosophy of the ancients. But we do not complain of this. Logic receives a fair share of attention in our colleges. In almost all of them it is a required study for at least one term; while the larger institutions offer advanced courses in theories of knowledge and belief.

This is all that could be expected. A pretty thorough indoctrination in logic can be effected in connection with forty or fifty class exercises; and half as many might suffice for imparting the rudiments. Or may we say that the minimum of required work should include not less than thirty recitations, or class-exercises; after which the young men might be left to their own election as to the further prosecution of this study?

So far as we know, Logic is never taught without the help of a text-book; though professors differ in the degree of their reliance upon this aid. The writer, who has used successively a considerable number of books, has always found it advantageous to select, with some freedom, the more important chapters, as subjects of recitation; and has supplemented the instruction thus given by a few informal lectures, and by some work required of every member of the class. He has also, of course, encouraged the students to read more than was imperatively prescribed.

He proposes now — so far as there may be need — to deal with his own book as he has dealt with those of others. For the chapters of the "Modalist" are of such a construction as to facilitate this method of procedure. They will be found to have so much independence of one another, that almost any of them could be omitted while the rest would remain comprehensible. And this is especially the case with certain chapters, such as the twenty-first, the twenty-second, and the twenty-third; in which the principles of the new analytic are somewhat minutely expounded. We think that a serviceable knowledge of inferences and syllogisms — so far as these are considered in existing manuals — can be obtained from chapters preceding and following those just mentioned.

Moreover, it will be noticed that the closing sections of several of the longer chapters are devoted to supplementary discussions; such as are consigned to small type in the author's metaphysical text-book.[1] In the present work this device, always unseemly, has not been thought necessary. The author is confident that any fellow-teacher who may honor him by employing the new logic as a means of class instruction, will sympathize with it sufficiently not to need specific directions concerning the use of it. Besides, every qualified professor can judge, better than any one else can, what the limitations, and what the possibilities, of his work may be.

CLINTON, N. Y., Feb. 1, 1891.

[1] A volume entitled "Mental Science."

CONTENTS.

		PAGE
PREFATORY DISSERTATION.		1

CHAPTER

		PAGE
I.	LOGIC DEFINED	13
II.	BELIEF, OR CONVICTION	21
III.	LOGIC DIVIDED	26
IV.	ENTITIES AND CONCEPTIONS	31
V.	GENERAL AND INDIVIDUAL NOTIONS	37
VI.	PREDICATIVE NOTIONS; THE "CATEGORIES"	45
VII.	PREDICATIVE NOTIONS; THE "PREDICABLES"	52
VIII.	THE DEFINITION OF NOTIONS	61
IX.	LOGICAL DIVISION	71
X.	PROPOSITIONS AND PREDICATIONS	79
XI.	CATEGORICAL PREDICATIONS	89
XII.	THE ILLATIVE PROPOSITION	98
XIII.	INFERENTIAL SEQUENCE	107
XIV.	ORTHOLOGIC INFERENCE	118
XV.	HOMOLOGIC INFERENCE	130
XVI.	INDUCTIVE REASONING	136
XVII.	HYPOTHETICAL AND DISJUNCTIVE REASONINGS	150
XVIII.	PROBABLE INFERENCE	162
XIX.	THE OPPOSITION OF PROPOSITIONS	174
XX.	THE CONVERSION OF PREDICATIONS	190
XXI.	CONTINGENCY AND ITS CONVERSION	204
XXII.	SYLLOGISMS	222
XXIII.	SYLLOGISTIC MOODS	242

CHAPTER		PAGE
XXIV.	The Pure, or Dogmatic, Syllogism . . .	261
XXV.	The Reduction of Syllogisms . .	279
XXVI.	Fallacies	290
XXVII.	Fallacies in Catenate Inference	304
XXVIII.	Exterior Catenational Fallacies	318

LOGIC AT THE PRESENT TIME:

A

PREFATORY DISSERTATION.

One hundred years ago the philosopher of Koenigsberg, in the preface to the second edition of his "Kritik," declared that logic had not advanced a step since Aristotle, and was, in fact, a completed science. According to Kant, authors subsequent to Aristotle had added nothing to logic, but had disfigured the science by the introduction of topics foreign to it. "For," says Kant, "logic is a science which has for its aim nothing but the exposition and proof of the formal laws of all thought, whether it be *a priori* or empirical, whatever be its origin or its object, and whatever the difficulties which it encounters in the human mind."

But, during this nineteenth century, logical questions have been discussed more earnestly than ever before, and, at the present time, no department of speculative investigation attracts greater interest than that relating to the laws of rational conviction. Differences, moreover, still prevail concerning the fundamental doctrines of this science. The only parts of logic on which there is general agreement are certain forms and rules which have descended to us from Aristotle. The philosophy of conviction continues a subject of debate; for which reason we cannot allow that logic is a completed science. No science can be regarded as complete till its principles have been established.

The treatise now offered to the public is the result of long-continued studies which have had for their object to place the doctrines of logic on satisfactory foundations; and it would be false humility were the author to conceal his assurance that these studies have been successful. He claims to have completed a work which Aristotle left unfinished, and that, too, in a way which would be approved by this great thinker were he now living. For Aristotle's "Organon" does not pretend to set forth a perfected system. It is not a treatise in which unity and simplicity have been reached through the ultimate analysis and the final synthesis of the laws of thinking; it is a collection of books written independently of each other, and whose discussions — especially when they relate to forms of argument — elucidate specific operations rather than universal principles. The writings of Aristotle, in general, reveal little effort at unification; he aimed not so much to produce systems as to discover and present truth respecting important topics. Different discussions show different analyses of the same subject; his statements of related truths occasionally lack co-ordination; and he is content, at times, with primary and superficial generalizations. Therefore, with that lasting strength which results from conformity to the individual and the actual, his philosophy exhibits also the obscurities and difficulties and defects of uncompleted doctrines.

We are aware that the claim to have reconstructed logic and to have made it a thoroughly satisfactory science is a bold one, and not likely to be immediately allowed. Even though one acknowledge his indebtedness to preceding thinkers; without whose labors success would have been impossible; and though he represent himself, not as a master-thinker, but as a disciple who has been fortunate in his day and opportunities, the author's estimate of his work will be received by many with incredulity and by some with ridicule. Yet he knows what he has been enabled to do; he is certain that he has found the truth on every important point; and, with this confidence, he comes before the public, not at all assured of his immediate reception, but willing to wait, if need be, till his views shall be understood.

The reader who may desire to comprehend the spirit and aims of the treatise now submitted to his criticism, should consider some problems which exercise logicians at the present time.

First, they desire *a clear definition of their science*. Aristotle does not give any definition. That already quoted from Kant is the one commonly found in text-books. For instance, Sir William Hamilton says that "logic is the science of the formal laws of thought, or of the laws of thought as thought." This definition is unsatisfactory; it itself needs to be defined. One might object to it that logic does not have all thought for its subject, but only rational thought; and also, that logic deals with rational thought not simply as such, but as the instrument and vehicle of conviction. We may, however, accept Kant's definition, provided it be taken to signify that logic is the science of the formal — that is, the essential and necessary — laws of rational conviction. For logic is a science which could be used by the rational beings of any universe.

Again, the discussions of our day call for *a true determination of the sphere and scope of logic.* The Kantian limitation of the science to the formal laws of thought is correct, if it be understood rightly. Pure, formal, or general logic does not consider those modes of enquiry or rules of procedure which are peculiar to any specific sphere of existence or investigation. Nevertheless, though thus limited, logic aims to understand all those modes of mental action which reason must employ, under whatever constitution of things, in her pursuit of truth. But if this be so, not only "pure," but also "modal," propositions and reasonings should be considered. That is, those processes of thought which follow the relations of contingency and of necessity, as well as those which use only simple assertions respecting classes and portions of classes, should be studied by the logician. Especially contingent and probable, no less than apodeictic, conviction, must be discussed. For contingency and probability are not confined to a specific sphere of being any more than necessity and certainty. They belong to the nature of things, and must be found in any universe. The fact is that problematic inference, though setting forth what is not necessary, is itself as necessary an act of reason, and as truly governed by law, as the apodeictic judgment.

This was Aristotle's view. He devoted far more attention to "modal" assertions and reasonings than to the "pure." In proof of this it is to be noted that four chapters of that book, the "Prior Analytics," in which syllogistic moods are examined, treat of those moods which are "pure," while fifteen treat of those which are "modal." But these modal syllogisms have been neglected by logicians for centuries, and of late years all the teachings, both of Aristotle and of others, concerning contingent and necessary sequences, have been "formally expelled from the science." In the words of Professor Bowen of Harvard, "The whole doctrine of modality is now rightfully banished from pure logic," as pertaining "not to the form, but to the matter of thought." "Pure" propositions and "pure" syllogisms, only, are considered as lying within the province of the logician. Thus logic has been *simplified* by the summary process of amputating its more troublesome part. But the question arises, 'Can desirable simplicity be obtained by a method which is founded on error, which divorces things most intimately related, and which necessitates superficial and one-sided views?' Cannot the difficulties of the case be solved in some better way than this? Some more natural way?

In this connection the name "Modalist," which has been given to the following treatise, may be mentioned. It is intended to indicate that the re-introduction of modality is characteristic of the new logic. Other features may equal this in importance, but none other has so evidently modified the rules and formulæ of the science.

A third desideratum, in order to a clear and satisfactory logic, is *a sound system of metaphysics, or ontology.* He who would understand thought as employed in rational conviction must study thought in its relation to objects. For thought, by reason of its very nature, corresponds to the nature of things; and therefore, merely as expressive of this truth, every thought may be said to have objectivity, whether it have an object or not. Without this characteristic thought could not serve the purposes of knowledge, or be of any logical importance. Conceptions are of interest to the logician only so far as they may, or do, correspond with realities and set forth truth or falsehood about them. Correct thinking is that which has this correspondence, or which, in an hypothetical case, would have it if the antecedent supposed were a reality;

thinking is incorrect when it is wrongly assumed to have this correspondence. Such being the case, the forms and sequences of thought, so far as it is the instrument of conviction, must relate to the nature and laws of things, and should be studied in the light of this relation.

It cannot be said that these principles have been rejected by logicians; neither can it be said that they have been accepted and applied. Certainly their significance has not been realized by those who identify things with our thoughts of them, or who deny that thing are as we conceive them to be, or who say that logic is concerned with thought only and not with things also.

The philosophy from which the following chapters derive their force has been named PERCEPTIONALISM, because it maintains, from an analytic and theoretical point of view, that what men call their perceptions are true perceptions of those very things which they say that they perceive. This philosophy prizes highly the Aristotelian doctrine of "common sense," or "common perception," — κοινή αἴσθησις, — but differs from it in being a developed system. It was constructed throughout upon a critical investigation of human thought, but, it is to be hoped, with a more exact initial observation of data than seems to have attended the "Kritik" of the illustrious Koenigsberg professor.

The author ascribes his success — or what he regards as his success — and his confidence in it, to his metaphysical preparation. He cannot see how a satisfactory logic can be constructed except in connection with a sound ontology.

A fourth requisite to the science of rational conviction, and one more specific than those already noticed, is *an analytic understanding of the nature of simple judgment and of the knowledge of fact or truth.* For these are both modes of mental assertion, and if we use the term "judgment" in its wide logical sense, cognition, the initial act of knowledge, is simply that species of judgment which results in absolute and well-founded conviction.

Both knowledge and judgment are expressed by the "proposition" in its most general assertive use; and their radical nature is to be ascertained by an analysis of the simple assertive proposition. Logicians give different accounts of this. Aristotle says that "a proposition is a sentence in which one thing is affirmed or denied of another"; Locke, that it is a statement setting forth the agreement or disagreement of one idea with another; Kant, that judgment is the application of a higher conception (the predicate) to a lower conception (the subject). "For example," says Kant, "in the judgment 'all bodies are divisible,' our conception of divisible is applicable to various other conceptions; among these, however, it is here particularly applied to the conception of 'body.'"

All these definitions, and others which might be quoted, are unsatisfactory. Aristotle's, though the best, is superficial and specific when it should be analytic and universal. It does not apply to all judgments, but only to the most common class of assertive propositions. Moreover, it gives a logical division of these rather than a true definition. The correct doctrine affirms judgment to be *the mental assertion of the existence or of the non-existence of something;* and that there are two modes of judgment. For every assertive proposition is either a simple existential statement, in which the existence or non-existence of the subject is set forth, or it is a predication-proper — an inherential statement — in which the predicate is set forth as existing, or as non-existent, in its relation to some subject. Without this doctrine any account of the laws of belief and conviction must be extremely defective.

A fifth essential in logical science is *a thorough theory of inference and of illative judgment and assertion in general*. Aristotle discusses only that specific mode of sequence which he calls the syllogism. Modern writers for the most part distinguish "mediate" inference — that is, the Aristotelian syllogism — from "immediate" inference. The latter, they say, derives its life from the laws of identity and of contradiction; the former, from the dictum of Aristotle, or from some similar principle founded on the relation of the generic to the specific. In addition to this they speak of hypothetical inference as following the law of reason and consequent.

Nothing could be more superficial, more inadequate, more

confusing, than any such analysis. Here logic must be wholly reconstructed. The one universal law of inference, to which all others are subordinate, is that of Antecedent and Consequent. The formula "*hoc est; ergo illud est*" expresses the nature of every illative sequence, however simple its antecedent may be or however complex.

Then there are *two generic modes of inference*, the orthologic and the homologic. In the one of these a consequent is inferred from an antecedent without reference to any previous case of similar sequence; in the other we infer a similar consequent because we perceive a similar antecedent. In the one, following the ontological connections of the elements of entity, we form direct "intuitions" of things as ontologically related; in the other, we infer a consequent by reason of the recurrence of its antecedent, on the principle that like logical antecedents — whether ontological or cosmological — are invariably followed by like consequents. This law — the homologic principle — supports not only induction, but all principiation whatever. It is the basis of all reasoning either to, or from, or in the general; that is, it justifies such reasoning.

Further, *illative propositions* hold an important place in the philosophy of ratiocination. Such propositions, whether they be pure or modal, are modal in meaning, and really express inference. They must be contrasted with simple factual assertions, whether singular or general. The subject of an illative predication sets forth an antecedent; the predicate — or rather the predicate part of the assertion — sets forth the consequent. When Kant says "body is divisible," he expresses the general sequence that "if there is a body, it can be divided." When the statement, "some snakes are venomous," is used as a principle in reasoning, this proposition, though "pure" because factual in form, is illative in force, and really signifies, "if there be a snake, it may be venomous."

The doctrine of the illative — or inferential — proposition bears directly on that of the Aristotelian syllogism. For that syllogism is best explained as the combination of two illative propositions so as to produce a third.

Sixthly, as already suggested, there is need that *contingent as well as apodeictic inference should be analytically explained.* This has not yet been done, because the metaphysical grounds of contingency have never been accurately determined. Aristotle distinguishes contingency from necessity, and analyzes many reasonings from contingent premises, but he develops no theory concerning contingent sequence — its nature, its origin, its diverse modes, and its relations to sequences in possibility, probability, and necessity. Nor has any one else given this topic thorough treatment. It is to be allowed, only, that the subject of probability has been handled by modern mathematicians with great ability, and that, in this way, some light has been thrown on the theory of problematic sequence in general.

Logical possibility — possibility in the widest sense — is the basis of contingency and of probability. The law of this mode of sequence is that *a thing is possible when one or more of its conditions exists.* By condition we mean a necessary condition, a *sine qua non.* Space, time, and an adequate cause were conditions of the universe. Building materials, a builder and his tools, his plans and his remuneration and other inducements, are conditions of a house. Now whatever either is or contains a condition renders the thing conditioned possible, so far as that condition is concerned.

Conditions are either causal, or constitutive, or concomitant. The first of these enter into and compose the essential cause of a thing; the second constitute its nature; the third are its necessary attendants and consequents. Space, Time, and the Creator are free from causal conditions.

This doctrine of conditions is the key to the philosophy of problematic sequence; and explains apodeictic sequence also. For, whenever a thing exists, then each of its conditions *must* exist, as above, in the cases of the universe and the house. Moreover, though conditions, as such, do not necessitate, but are necessary, they may be said to have a necessitative tendency or value. For the core, or vitalizing part, of any ordinary logical necessitant is composed of necessary conditions of the consequent.

This core is the *exact* logical antecedent of the consequent, and may be called its necessitant condition, because it both necessitates and is necessary. Whenever an antecedent is constituted exclusively from necessary conditions, it reciprocates with its consequent, and may be inferred from it conversely. If either antecedent or consequent exist, the other must exist also; and if either be non-existent, the other must be non-existent. The occurrence of such reciprocations is especially noticeable in mathematical sequences.

A given collection of circumstances — or a case — may contain a set of conditions capable of being filled out *in any one, but in one only, of a limited number of ways so as to constitute a necessitating condition*. When we know that there will thus result an antecedent of necessity in some one way, and have no reason to suppose that this will occur in one way rather than in another, we call each of the possible consequents a *chance*, and we say that the chances are equal to one another; because we divide among them the confidence of certainty. Then, should a proportion of these chances support some general specific consequent, we say that this consequent has a probability expressed by the ratio of the chances for it to the whole number of chances. The probability that an odd number will turn up on one cast of a die is one-half, because three out of the six possible individual consequents favor an odd number.

Now, when we know only that a certain specific consequent is supported by chances and are unable to determine the ratio of the chances for it to the whole number or to those against it, then *the indeterminate probability thus arising is that contingency — or contingent sequence — of which logic treats*.

A most important modification of contingency takes place when it is *guarded against a necessity of the opposite*. This is effected either when the consequent asserted as contingent is known to have already sometimes accompanied the antecedent, or when the very nature of the antecedent is seen to preclude impossibility. "Man may be wise" is guarded against a necessity of the opposite because men *have been* wise. This renders it clear that further investigation will not show that man (as such) cannot be wise. So also ace *certainly may* turn up on the cast of a die, because there is nothing in the nature of a die or in the act of throwing it to prevent ace appearing.

Another mode of contingency which is *not* guarded is frequently used by the mind. But guarded contingency is that assumed in the modal syllogisms of Aristotle, and has a just pre-eminence.

The theory of sequence based on th Doctrine of Conditions renders possible a simple and intelligible account of modal reasoning, and, indeed, makes the modal syllogism, in its apodeictic and its contingent moods, the syllogism *par excellence*, and that to which all syllogizing must be referred. Thus, too, what has long been the terror and despair of scholars has been converted into the crowning part of logical science.

Once more, and in the seventh place, we may say that *the Aristotelian syllogism calls for more accurate definition than it has yet received.* The ordinary description of it as the "mediate inference" is indefinite and unsatisfactory, and the explanation of its process as the comparison of two terms, or two ideas, with a third so as to determine their relation to each other, is a vague, inadequate attempt at the expression of truth. Aristotle himself says that "a syllogism is a sentence in which, certain things being laid down, something else different from those things necessarily follows by reason of their existence." This, also, is superficial and inadequate. For the questions arise, 'What is the nature of the things laid down? and what is the nature and ground of the sequence?' There is a lack here similar to that in Aristotle's definition of the proposition.

The only mode of inference which possesses all the "accidents" of syllogistic figure and mood — the only style of sequence to which all the rules of syllogizing apply — is *that which combines two general illative propositions so as to produce a third:* it is the process which obeys the law that the antecedent of a second antecedent is antecedent also of the second consequent; or (from another point of view) the law that the consequent of a prior consequent is consequent also of the antecedent of that prior consequent, and is therefore a "consequent-consequent."

We cannot now speak further of this law or of its relations to other principles of inference. Our present object is not to expound the new doctrines, but only to indicate their nature. We must, however, add that no change has been proposed in those forms and rules of syllogizing which have come down to us from ancient times, except in the way of unification and of a slight addition. After the nineteen commonly recognized syllogistic moods have been interpreted by modal laws, twelve other moods, very simple in structure, have been added so as to express conjectural, or unguarded, sequences in contingency. Thus every mode of syllogizing used by the mind has been provided for. These unguarded moods are equal in philosophical, though not in dialectic, importance to those ordinarily allowed; they have been neglected heretofore. Some of them indicate methods of reasoning which are quite common.

One result of the new analysis has been to exalt the general doctrine of inference and its modes above that of the Aristotelian syllogism. The sphere of the latter has been restricted to "general catenate inference"; while other specific modes of sequence have been assigned distinctive places. This but carries out a tendency in modern logic, according to which various modes of inference have been treated independently of the syllogism and as following principles of their own.

Even yet, however, logicians do not make sufficient allowance for the fact that different modes of inference may employ the same linguistic expression. A verbal, or superficial, form of sequence should not be taken as the ultimate explanation of inferences essentially diverse. Logical theory should not rest in the secondary and the ministerial, but should point directly to the ultimate. Following this rule, the Aristotelian syllogism will be given a true pre-eminence, yet also a definite and limited place, among modes of inference.

In the foregoing remarks no enumeration of new doctrines has been attempted. This preface is intended merely to show the spirit and aim of the treatise which it introduces. Possibly other teachings of the book may seem to some of greater interest than those already referred to. But this may be said regarding every doctrinal modification: it has been introduced without any love for novelty, and only under a sense of necessity, and with a profound confidence in that underlying system of philosophy which has suggested the innovation. The positions taken relating to logical definition and division — to the categories and the predicables of Aristotle — to induction — and concerning probable judgment — the specific laws of orthologic sequence, whether mathematical or metaphysical — the specific modes of homologic sequence — the quantification of terms in propositions — the opposition and conversion of predications — and concerning fallacies and their classification, have all been controlled and determined by the analysis of Perceptionalism.

We would say, in conclusion, that, in one respect at least, the aim of the present work has been very limited. The history of opinions and the discussion of views which are well worthy of attention, have been quite beyond its scope. The endeavor has been simply to elaborate fundamentals. Perhaps, after a time, some additional chapters may be composed in criticism of important theories and in further elucidation of the system now submitted.

HAMILTON COLLEGE, CLINTON, N.Y.
Nov. 4, 1890.

THE MODALIST:

OR,

THE LAWS OF RATIONAL CONVICTION.

CHAPTER I.

LOGIC DEFINED.

1. Origin of the name. 2. Not the science of "thought as thought," nor of "inference" only. 3. The science of rational conviction. 4. Reason not radically different from lower faculties, but a special endowment and development. 5. "Discursive" reason is articulate and intentional; "intuitive" reason, habitual and instantaneous. 6. Truth is (*a*) attributal, (*b*) objectual, (*c*) subjectual, or propositional. The term "subject."

1. THE name "Logic" was originally the Greek adjective corresponding to the noun λόγος, which noun signifies either language or that rational and elaborated thought of which language is the expression. As descriptive of a science the adjective λογικός was employed either in the singular or in the plural. The plural phrase, τὰ λογικά, might be translated "the principles of rational thought." It sets forth the science as composed of parts. It is similar in origin to the expression, τὰ μεταφυσικά, or "metaphysics," a name anciently given to the philosophy of the ultimate in conception and in existence.

The singular designation, "ἡ λογική," is that Anglicized by the word "logic." The meaning of it is fully expressed in the original language by adding to it a noun, either ἐπιστήμη or τέχνη; and the phrase thus formed may be translated "the science, or the art, of rational thinking." The term τέχνη with the Greeks, like the term "art" with us, is often used to designate a practical science. Logic even yet is sometimes called an art because it not merely elucidates truth, but also formulates rules and gives useful directions.

2. This science has been variously defined by modern writers. Most of them say that "Logic is the science of the laws of thought," and some make this statement emphatic by saying "of thought as thought"; that is, of thought considered simply, or chiefly, as to its own nature. One great objection to this definition, at least as it is commonly given, is that it uses the word "thought" in a narrow and technical sense without sufficient explanation. This word is applicable to all our thinkings as well as to those exercised in connection with rational conviction. Memory, imagination, sense-perception, consciousness, have each its own mode of thought; to say that logic is the science of thought without showing clearly what kind of thought does not satisfy the enquiry of the mind.

But a second and more serious fault in the above-mentioned definition is that it tends to conceal a radical distinction of mental science, namely, *the distinction between thought, or conception, and belief, or conviction*. This tendency is especially noticeable when we are told that Logic is the science of thought as thought. For, according to the most natural use of terms, Logic does not consider thought simply as thought, but thought always and only as the instrument and vehicle of conviction; and the laws of thought as the organ of belief cannot be clearly understood if we do not first recognize the distinction between thought and belief.

Again, some have defined Logic as the science of reasoning, or inference. For instance, Professor De Morgan calls it "the calculus of inference necessary and probable." This definition is not sufficiently broad; Logic discusses not only inferences and reasonings, but also conceptions, or notions, and statements, or propositions. Nor are these considered merely in subordination to reasonings, but also as having an independent use of their own. An important part of Logic aims simply to render our notions and statements more adequate and efficient as the embodiments of truth.

Another class of writers, desiring to emphasize the practical office of Logic, say that it is the science which teaches "the right use of reason." This definition cannot be greatly condemned, yet is wanting in completeness. The rules and directions of Logic as an art cannot profitably be separated from the philosophy of our rational operations. To say merely that Logic teaches the right use of reason is not a sufficient recognition of that scientific spirit without which any logical system would be weak and lifeless. Moreover, the term "reason," when used without qualification, covers a wider ground than is surveyed in logical discussions. In particular, that constructive imagination which produces poems and works of fiction is something pre-eminently rational; yet it does not fall within the cognizance of the logician.

3. Perhaps the definiteness of conception for which we have been seeking may be obtained in connection with the following statement: *Logic is the science of the operations and products of the rational faculty in the pursuit and use of truth.*

The distinction, incidentally assumed in these words, between the operations and the products of the mind is worthy of some attention, because it differs from that existing between material products and the labors in which they originate. In the latter case the things distinguished are of totally diverse natures, whereas the mental conception or conclusion, which results from some rational process, is a thing of essentially the same kind with the steps which lead to it. It is simply a completed thought or conviction which the memory retains, and which the mind can recall and use. Hence, after a mental product has been formed, it may immediately become part of a process which aims at a further product; as, for instance, when one notion, after being defined, is employed in the definition of another.

But the important point in our conception of Logic is, that this science considers reason, or the rational faculty, only so far as it is engaged in the pursuit and use of truth. On this account, if brevity were desired, it might be sufficient to say that *Logic is the science of rational conviction;* for belief, or conviction, is always the apprehension by the mind of something as true. That such is the essential character of Logic will be evident to any one who may examine the various systems of doctrine which have gone under this name.

At the same time Logic is not the science of conviction in general, but only of those modes of conviction which depend on the exercise of the reason. Those cognitional convictions which are not of a rational origin, though recognized by the logician, have only a preparatory and subordinate place in his discussions. One knows, from consciousness, of the pleasure experienced in meeting with a friend, or, from sense-perception, of the size, weight, solidity, roughness, and coldness of a stone; but such cognitions are not exercises of the reason, and are not investigated by the logician. This is true also of those perceptions of times, distances, changes, and relations which accompany the operation of sense-perception and consciousness. Without any rational process, a person holding two stones, one in each hand, would know that they exist contemporaneously, that they are separate in space, that they are similar to one another, that the one is heavier or rougher than the other, and so on. In short, all presentational cognitions, and the memories consequent upon them, are presupposed or taken for granted, in the science of rational conviction.

In speaking of Logic as a science, we would not ignore those practical aims, on account of which it has been called an art. Some authors have discussed logical questions in a purely theoretical spirit and without any attempt at useful directions. In this they deviate from that conception of the science which the experience of past times has shown to be both reasonable and advantageous.

4. Accepting the definition that Logic is the science which discusses the operations of reason in the pursuit of truth, let us consider attentively two leading ideas contained in it; let us determine exactly what we mean by reason and truth.

Reason, or, as it is sometimes called, the rational faculty, is a development of intellectual power which, because of its great importance and wonderful accomplishments, is distinctly noticed and named; yet it is not a faculty radically different in nature from our lower mental capabilities. A man of genius differs from his fellow-men, not in the nature of his gifts, but in the natural strength of them and in the degree of their development. In like manner, reason is to be distinguished from those powers of mind which man has in common with the more intelligent brutes rather as a special endowment of strength than as a faculty of a distinct nature. When Locke speaks of "that faculty whereby man is supposed to be distinguished from the beasts, and wherein it is evident that he much surpasses them," his words must be received with care. Examination shows that reason is not a faculty separate in its nature from our other powers, but only a special endowment of intellectual ability. The perceptions of sense, no less than those of the rational faculty, employ notions, judgments, and inferences; but reason far transcends all sense-perceptions in the grasp of her apprehension and understanding. In like manner, there is an exercise of the faculty of reasoning, or ratiocination, which falls far short of what we call reason. Many brutes exhibit some power even of connected reasoning. Reason is that gift by which man is capable of language, of civilization, of material social and intellectual progress, of civil government and laws, and of moral and religious life.

The superiority of this endowment to the lower powers of mind is manifested principally in two particulars. In the first place, rational *conceptions* are peculiarly *comprehensive ;* and secondly, resulting in part from this comprehensiveness of conception, rational *judgments* are peculiarly *penetrative.* Reason can seize and hold under consideration many things at once, so as to consider fully their nature and relations; and, while doing so, she reaches a knowledge of things which are invisible to lower powers of thought. So far as sense-perception is concerned, a brute sees the different parts of a locomotive as well as a man; but no brute can understand the relations, use, and value of each part, and by what process the whole contrivance accomplishes its work. Rational intelligence not only perceives these things, but constructed a locomotive in thought before such an invention ever existed.

Philosophers agree that, in the human mind at least, reason is exercised in two modes, the intuitive and the discursive, but they differ concerning the way in which these modes of reason are related to one another. Some hold that rational intuition is entirely without a process, or, at all events, wholly different in nature from rational discourse. The better opinion is that the intuition of reason is an instantaneous action the rapidity of which, resulting from the habitual and spontaneous use of certain modes of apprehension, causes the steps of the process to escape detection. Believing this, we must hold the intuitive reason to be a faculty of a very different nature from that power of "intuition" by which necessary relations are immediately perceived, and which enters as an element into every phase of human cognition.

The discursive mode of reason is that ordinarily employed in all our deliberate investigations. It is distinguished from the intuitive by being more analytical, articulate, and consecutive, and in being immediately under the guidance of the will. This form of the faculty, also, is the proper subject of logical principles and rules, because it alone admits of direct self-inspection and regulation. Yet an understanding of "the discourse of reason" enables us to understand "the intuition of reason," as well; the two being radically of the same nature. The rapid mode of reason may be compared to that motion of spinning or weaving machinery which is too swift for observation: the more deliberate mode may be likened to the working of a type-writer or a telegraphic instrument, every movement of which is an intentional act of the operator. The intuitive mode becomes understood when the same conclusions to which it comes quickly are reached by the consciously directed methods of mental discourse.

5. The question, "What is truth?" was often asked by ancient philosophers, and with them it mostly had a moral significance and meant, "What is the true end of life?" The first aim of the thinkers of antiquity was to find some essential principle the knowledge and observance of which might lead men to true happiness. In modern discussions, the term "truth" is more commonly used in that primary and literal sense which it has when we say that a statement is true, or is a truth, and deny that it is false. The truth thus mentioned has been called intellectual truth, and has been distinguished

in this way from that more specific kind which is ethical or moral. For truth in general and by reason of its essential nature is closely related to intellect.

This intellectual truth is of three modes, or denominations, which are intimately connected with one another. First, there is *attributal truth*. This is that defined by St. Thomas Aquinas when he says, "The truth of thought is a correspondence of thought and fact according to which thought says that what is, is, or that what is not, is not." (Veritas intellectus est adaequatio intellectus et rei, secundum quod intellectus dicit esse quod est, vel non esse quod non est.) Evidently if a statement — for example, that "the man is rich" — be true, there is a fact existing outside of one's thought, and also a proposition within the mind corresponding to the fact; and the truth which we ascribe, or attribute, to the proposition, lies in this correspondence.

Again, there is *objectual* truth. This is not any correspondence, but it is the fact, or reality, which is the object of the mind's knowledge, and which corresponds to the proposition in the mind. Accordingly we sometimes say, "That is the truth," our meaning being, "That is the fact." In such language fact, as the basis and object of knowledge, is called truth.

Finally, there is *subjectual, or propositional*, truth. The term "subject," when opposed to the term "object" in modern philosophy, signifies the mind as the subject of impressions from objects and of ideas about them. Subjectual truth, accordingly, is the ideas or conceptions of the mind considered as corresponding with facts or objects known. This may also be styled propositional truth, because when expressed fully it assumes the form of the assertive proposition.

For belief, or conviction, cannot be exercised on the mere conception of a thing as to its nature, however correct and complete this conception may be. There is always need that we should conceive of a thing as existing or as non-existent. To believe in God is to believe in the existence of God, or in the proposition that God exists; to believe in the justice of

God is to believe in the existence of His justice, or in the proposition that God is just; and to disbelieve in God and His justice is to believe that they do not exist. It is because assertive propositions set forth things either as existent or as non-existent that they are naturally fitted to express subjectual truth.

The signification of the noun "subject," referred to above in connection with the adjective "subjectual," belongs chiefly to the discussions of psychology. It is to be distinguished from the ordinary meaning of this word in Logic, according to which it is opposed, not to the term "object," but to the term "predicate." In the distinctively logical sense a subject is anything whatever of which anything may be affirmed or denied. But the doctrine of truth pertains to philosophy in general, not to Logic only; and therefore we need not confine ourselves, in the statement of it, to strictly logical terms.

When we say that Logic considers the operations of the reason in the pursuit and use of truth, it is clear that the reference is to subjectual, or propositional, truth. This is that which the mind immediately apprehends and employs; it is only by obtaining possession of this that the mind becomes sensibly related to attributal and objectual truth.

CHAPTER II.

BELIEF, OR CONVICTION.

1. The two primary powers of mind, — thought, or conception, and belief, or conviction. 2. Belief and knowledge defined. 3. Judgment and cognition defined. 4. Inferential judgment, (*a*) either apodeictic or problematic, (*b*) either actualistic or hypothetical. 5. The sphere of general, or "pure," logic.

THOUGHT, or conception, and belief, or conviction, may be termed the primary powers of the intellect, because, in their exercise, the work of mind is directly accomplished: our other powers, such as attention, association, abstraction, generalization, synthesis, and analysis, are secondary, because their function is to modify the operation of thought and belief.

1. Of the two primary powers, thought is the more prominent in our experience; for belief is felt only as an accompaniment of thought. We may have conceptions unattended by convictions, but we cannot have a conviction except as attached to some conception. Moreover, in every enquiry respecting belief, questions respecting the origin and mutual connections of our thoughts are implicated. This close association of belief with thought has led many writers to treat belief as if it were merely a peculiar, or, it may be, a superior, kind of thought. This is a mistake, and the cause of wide-spreading confusion. President McCosh ("Scottish Philosophy," p. 384) says truly, "Belief should have a separate place in every system of psychology"; to which we add, "and in every system of logic also."

2. But, before proceeding farther, we must remark that, in the present discussion, the term "belief" is used in a very wide sense. Ordinarily belief signifies a mode of mental confidence which falls short of knowledge, yet which is greater than mere guess-work or presumption. Seeing certain weather indications, one might say, "I believe, though I do not know,

that it is going to rain." We now include under belief every degree of confidence respecting the truth of a thing from the weakest conjecture to the most absolute assurance. According to this signification knowledge is a kind of belief; for knowledge is absolute and well-founded certainty.

At present, also, we use the term "conviction" as synonymous with "belief," though conviction strictly indicates belief, not simply, but as founded on evidence. In like manner, we employ the terms "conception" and "thought" interchangeably, though a conception properly signifies a thought formed synthetically.

The most important point in the doctrine of belief is, not that conviction takes place only in connection with conception, but that belief is possible only when the thought of existence or that of non-existence is united with or included in our conception of a thing. This truth has been expressed too strongly by those who say that belief takes place only in connection with propositions. It is the essential and formal function of propositions to set forth things as existent or as non-existent, but any notion may become matter of belief if it only be an existential conception; that is, if it have, as one of its elements, the thought of existence or that of non-existence, whether this element be prominent in our conception or not. For instance, should one predicate something respecting an existing object, saying, "My friend is faithful," the subject-notion, "my friend," presents the object as existing, and as believed in, though the existence directly asserted by the proposition is not that of the friend, but of his faithfulness.

3. The same necessity which leads to a wide use of the term "belief" calls for an equally broad use of the term "judgment"; for judgment is the initial act of which belief is the permanent and reproducible product. Ordinarily judgment signifies the formation on evidence of a probable conviction. Hence Locke says, "The faculty which God has given to man to supply the want of clear and certain knowledge is judgment, whereby the mind . . . takes any proposition to be true or false without perceiving demonstrative evidence in the proof." But logicians have found it advantageous to give the name "judg-

ment" to the assertive faculty in general; in other words, to that faculty, in the exercise of which we form convictions of any kind, and are led to embody these convictions in propositions or statements. According to this use of language cognition, the initial act of knowledge, is a mode of judgment, knowledge being, as we have seen, a mode of belief.

If we consider our convictions and the judgments productive of them with reference to their primary origin and mode of formation, they may be divided into two classes, — the presentational and the inferential. The former of these includes our cognitions of such things and relations as are immediately present to the soul in space and time; and with these cognitions we may also classify, as things of the same logical relations, the simple reproductions of presentational perceptions. Our first perceptions are important because they are the basis of all subsequent knowledge and belief, but the special consideration of them belongs to psychology. They furnish those materials of fact which reason uses, but are not themselves distinctively rational. While the logician recognizes them, he does not make them the subjects of his investigation.

Inferential convictions are those which assert the existence or the non-existence of things not immediately present to the soul. It is with them that the discussions of logic are chiefly occupied. They differ from presentational cognitions in that the latter do not depend on any previous knowledge, while inference assumes something as already known to be fact, and then asserts some second thing as a fact connected with the first.

4. Considered with reference to their own nature and operation, inferential judgments are divisible into two principal classes, — *the apodeictic, or demonstrative, and the problematic, or contingent.*

The apodeictic inference leads to an absolutely certain conclusion, and excludes the possibility of a thing being otherwise than as it is shown to be. Such are mathematical demonstrations and all reasonings which infer things as necessarily related to given fact. When a surveyor knows the length of the sides of a field and the angular measurements of its corners, he calculates the area by an apodeictic, or demonstrative, process.

Problematic inference is based on the consideration of things as possible or as contingent, and produces forms of conviction weaker than those which result from demonstration. Contingency is a mode of sequence approaching probability: it is an expectant possibility. It arises when an antecedent of possibility admits only a limited number of possible consequents, some one of which must be realized. Old age is one of several conditions, one or other of which must belong to every man. Therefore it is contingent to man to be old.

Contingency is best discussed as a mode of possibility which prepares for probability. Many, following Aristotle, and neglecting the distinction between contingency and probability, treat both modes of sequence under the head either of contingent or of probable inference; but a wise use of terms limits "contingency" to those cases in which a thing is looked for, or in any degree expected, as possible, without having its probability determined, and limits "probability" to those cases in which some *proportion* out of a total number of chances is found or estimated to favor some conclusion. Thus it would be a judgment of contingency to say, "A merchant may prosper, and become wealthy"; but of probability to say, "The wise and prudent merchant will prosper." Contingency lies between possibility and probability, being more than the one and less than the other. It passes into probability whenever the ratio of the chances is estimated. Both contingency and probability *expect*, which accounts for their being often included under the general name "contingency"; but they are clearly distinguishable.

Another division of inferential judgments separates them into *the actualistic and the hypothetical.* This distinction relates not so much to the internal nature and operation of inferences as to the character of the grounds on which they are based, and of the convictions which they produce. For when an inference, whether apodeictic or problematic, arises from our knowledge of fact or from belief in what we take to be fact, the conclusion of it asserts fact, or at least the possibility or probability of fact; and the inference is actualistic. But if our reasoning be based on supposition or assumption, the conclusion sets forth only what would be fact (necessarily, or possibly, or probably), provided the supposition were realized. In this case the inference is hypothetical, and asserts what, in the most literal sense, may not be true at all. Should we suppose one of the Green Mountains to be of solid gold, we might assert Vermont to be the wealthiest State in the Union, and the inference would be correct; yet evidently neither premise nor conclusion would set forth reality.

Hypothetical inferences may be based on antecedents to which no facts ever correspond, but more frequently they present the abstract operation of some law of existence or of nature. For it is only by an exercise of the imagination that we can conceive of the separate working of a law which never is seen to operate except under a complication of modifying circumstances. Hence hypothetical inferences are largely employed in science.

5. Some writers teach that neither the inference of the actual nor that of the probable or of the contingent lies within the sphere of logic. Rightly conceiving of logic as the general science of our rational operations and as independent of any particular branch of knowledge, they say that the theory either of problematic or of actualistic conviction is necessarily connected with that knowledge of specific classes of things which experience gives us, and that the logic of hypothetical demonstration, alone, is an abstract and ontological science.

These views are not well founded. While all the methods of reason should be illustrated and tested by their application to particular cases, the principles of actualistic conviction are not specially connected with any one class of facts or objects, and those of problematic inference are such as must govern finite intellects in their judgments relating to any universe, or system of affairs, in which they can be placed. If the subject of logic as a general science — of "Pure Logic," as it has sometimes been called — be rational conviction in general, then logic must consider actualistic as well as hypothetical, and problematic as well as apodeictic, inference. All these modes of rational conviction, together with their principal varieties, are such as must be followed, by minds like ours, in any universe, or system of things, whatever.

CHAPTER III.

LOGIC DIVIDED.

1. Logic is objective or subjective. 2. Is general, or abstract, and special, or applied. 3. Is "pure," or "formal," and mixed, or modified — but ambiguously. 4. The terms "directive" and "corrective" proposed. 5. Logic concerns (*a*) notions, or conceptions, (*b*) judgments, or assertions, (*c*) inferences, or reasonings.

In order to render our conceptions of logic and of the sphere of its instructions more definite, various distinctions and divisions have been made.

1. First, *objective has been distinguished from subjective logic*, or, in the language of the schools, Logica Systematica from Logica Habitualis. The necessity for this distinction arises from the double signification of the word "art." Since this word may indicate either a system of practical principles or an acquired facility in the application of those principles, there are two senses in which one may be proficient in logic. He may be a theoretical logician, well-acquainted with the laws and rules of thought, or he may be a practical logician, skilful in the application of the rules. While habitual logic is a chief end of systematic logic, these two "arts" are distinct acquirements, and do not always accompany one another. He who would be in every sense a complete logician must not merely familiarize himself with the principles of correct thinking, but must also sedulously practise them. Nor should he expect to obtain from books, or even from instructors, much more than a useful knowledge of right methods.

The foregoing distinction has sometimes been called a division of logic. But it does not really divide the science. It only explains how the term "logic" may be employed in a secondary sense. Subjective and objective logic cannot naturally be regarded as parts of the same whole; and the logic set forth in books, which is that commonly spoken of, is wholly objective.

2. Again, *general, or abstract, logic has been distinguished from special, or applied, logic.*

Every department of enquiry is properly subject to various regulative principles connected with the specific character of its investigations; and these principles, though immediately subordinate to the universal rules of right thinking, constitute a separate system of directions. Mathematical progress is promoted by a knowledge of the correct use of diagrams, instruments, figures, symbols, modes of notation, and methods of calculation. In courts of law barristers and judges are governed by rules respecting the pertinency and value of different modes of proof and the fair interpretation of legislative enactments. The theologian appeals to the canons of Biblical exegesis; and the psychologist, who would ascertain the laws of mental life, first determines on what sources of knowledge and on what methods of enquiry he may rely. In short, every science has its own principles of procedure, which, as supplementary to the rules of right thinking in general, may be called the special logic of that science.

But the several regulative codes now described are no part of logic in the ordinary acceptation of the word; for by "logic" we commonly mean that general science which sets forth those forms and laws which rational conviction should observe, no matter what may be the specific nature of the topics considered. The distinction between general and special logic is not properly a division of that general science. Each special logic involves considerable acquaintance with the department of investigation to which it pertains, and is simply that philosophical "introduction," or "methodology," without which great progress can scarcely be hoped for in any branch of knowledge. Every such code is a valuable addition to the science which it is intended to promote, and should be studied as a part of that science.

3. Again, *pure, or formal, has been contrasted with mixed, or modified, logic;* though logicians differ greatly in their explanations of this distinction.

Some say that general, or abstract, logic is "pure," because unmixed with the principles of any specific science, and "formal," because it sets forth the radical methods employed by reason in every sphere of enquiry; while particular methodologies modify the general rules of reasoning by mingling their own directions with them, and therefore constitute mixed, or modified, logic. In other words, Pure, or Formal, Logic is just the same as General, or Abstract, and Mixed, or Modified, Logic is just the same as Special, or Applied. This use of language is quite common, and is so supported by authority that it cannot be condemned or avoided; yet it is really undesirable. It repeats a distinction already provided for, and, as we shall see, conflicts with another and better use of terms.

Again, those who hold that the "necessary" laws of thought pertain only to hypothetical demonstration, confine the terms "pure" and "formal" to apodeictic logic, and relegate to mixed logic the consideration of actualistic conviction, of probability and contingency, of doubt, and of error. This division of the science and the implications of it cannot be allowed. The theory of demonstration cannot be separated in this way from the rest of logic. The same immutable and ontological laws underlie all modes of sound judgment and correct inference.

According to a third method of employing the terms in question, the logic of correct conviction is called "Pure," or "Formal," and that of imperfect and erroneous thinking, Mixed, or Modified. We can conceive of a purely intellectual being, unaffected by any cause of error, and compare him with creatures like ourselves who are subject to mistakes. And our mental action, so far as free from failure or delusion, might be held to obey the laws governing that pure intelligence; while our deviations and delinquencies in the pursuit of truth would be accounted for by influences which mingle with our thinkings and lead them astray. Hence we discriminate between the philosophy of the defective use of reason and that of correct and normal thinking. The distinction thus made is a true division of General, or Abstract, Logic.

4. At the same time, since logicians have disagreed in their

use of terms, two new names may be of service here. Were we, instead of the last distinction, to designate Pure Logic as Directive, and Modified Logic as Corrective, and were we to assign to the one the perfect and normal modes of rational conviction, and to the other the imperfect and abnormal modes, all room for misapprehension would be taken away.

But, while dividing logic into the Pure, or Formal, or Directive, and the Mixed, or Modified, or Corrective, we do not mean to say that the discussion of correct and that of incorrect processes should be wholly separated from one another. Clearness of statement and an orderly arrangement of details may require some separation, but we must not lose that advantage which accrues from the immediate contrast of perfection and imperfection. The division of logic into the Directive and the Corrective is principally significant as marking two lines of thought which run parallel with each other in logical investigations.

5. The distinction between actualistic and hypothetical conviction, though fundamental in logic, does not yield any division of the science. The difference of these modes of belief, both as to nature and origin, is very apparent, and the forms and processes of thought in connection with which they are experienced are perfectly similar. To determine whether a conclusion be actualistic or hypothetical, we have only to know whether it be drawn from fact or from supposition. This distinction, therefore, does not give rise to any great variety of discussions.

But an important division of logic is based on those three radical modifications of mental action which reason employs. For every exercise of rational thought is either a conception, or a judgment, or an inference; and every question in logic concerns one or other of these three things. The necessity of grouping according to this division soon becomes evident to the investigator, and it is also perceived that there is a natural order of succession for them, namely, that conceptions should be studied before judgments, and judgments before inferences. Hence most text-books contain three principal parts, corresponding to these three general topics.

But here we must remark that the logical division of a body of scientific knowledge should not be confounded with the orderly plan of a treatise; though these things often go by the same name. The object of logical divisions is to impress upon us certain pervasive and fruitful distinctions; the arrangement of a treatise is designed to facilitate our progress in the understanding of doctrines. Accordingly, in a scientific book, several radical divisions may be given, while only one arrangement of topics can rightly be adopted. From the nature of the case, indeed, any wise order of discussion must refer more or less directly to logical division, but the work of arrangement should not be so controlled by this relation as to be prevented from the free pursuit of its own proper aim.

These remarks may be illustrated by the plan of procedure chosen for the present treatise. It is essentially the same with that commonly adopted. It is based on the division of our rational states into conceptions, judgments, and inferences, and also on the fact that the doctrine of inference calls for a considerable variety of discussions, and occupies an extended place in logic.

Having now finished some necessary introductory dissertations we shall apply ourselves, in the next part of this treatise, to questions concerning conceptions, or notions. Then we shall take up judgments, or assertions. After that we shall discuss the radical laws and forms of inference; whether they belong to the apodeictic (or demonstrative), or to the problematic (or contingent) inference. This will prepare us for the composition of inferences and the conclusions thereby obtainable; which things fall under the head of syllogisms. Finally, some closing chapters may be specially devoted to fallacies and the causes of error.

CHAPTER IV.

ENTITIES AND CONCEPTIONS.

1. Entities, or objects, and notions, or conceptions. 2. Objectivity and objectuality. Truth and error. 3. Positive and negative (*a*) facts, (*b*) notions, (*c*) convictions. 4. Schematic conceptions. 5. Categorematic and syncategorematic words. 6. Subject and predicate. Substance and Accident, or *Substantum* and *Ascriptum*.

1. AN entity is anything whatever that does, or may, exist. Spaces, times, substances, powers, actions, changes, quantities, and relations, are so many kinds of entity. Whatever actually exists is a real entity; and when a thing does not exist, but is merely conceived of as existing, we use similar language to that which we would employ if it existed, and say that it is a possible, or an imaginary, entity. In the strictest sense that only is an entity which really exists. The essence of entity, however, does not lie in its existence, but in *its being that which exists*, and which, therefore, also may be of this or that nature.

The word "entity" is equivalent to the word "thing" in that wide sense according to which we speak of all beings, or existences, whatever, as things. The advantage of the philosophical term is that it has one signification only, while the word "thing" has many meanings.

That action or state of intellect which corresponds to any entity is called a notion, or conception; the entity of which we conceive is called the "object" of the conception, and the conception, as related to and corresponding with its object, may be said to be objective, or to have objectivity.

This objectivity belongs to the essence of thought. Any psychical activity which does not correspond to things, or entities, is not thought, but some other form of experience. To this statement, however, the thoughts of existence and of non-existence, and they alone, present an exception. Existence and non-existence are not things, or objects, in the full sense of these terms, though they may be thought of just as things are thought of, and must be allowed (that is, in all cases of fact) to have a kind of objectuality.

2. By "objectuality" we mean the character of things as being actually or possibly correspondent to our thought. The objectuality of entity is the counterpart of the objectivity of conception. But this objectivity of thought and this objectuality of things do not involve that a thought and the entity corresponding to it are of the same nature, or that they resemble one another, or that, if either exist, the other must exist also. They only imply that the nature of the one corresponds with the nature of the other. If the existence of a conception always involved that of the corresponding entity, there could be no such things as truth and error. Truth lies in the conformity of thought with fact, or with what, in case some hypothesis were realized, would be fact; while error is the disagreement, or want of correspondence, between thought and fact.

3. Now fact is of two kinds or modes, — the positive and the negative. According to the first the existence of a thing is a fact; according to the second, the non-existence. It is as much a fact that there is no bread in the cupboard, when that is true, as that there is bread in the cupboard, when that is true. Consequently, and corresponding to the positive and negative modes of fact, there are two modes of conception, — the positive and the negative.

These are expressed, respectively, by such terms as "bread," "a loaf," and "no bread," "no loaf." Commonly a thing is set forth as existing by a noun without the adjective "no," and as non-existent by the same noun with the word "no" prefixed. At first sight it appears self-contradictory to speak of a thing, or entity, as non-existent; and it would be so if we intended to speak of a real entity as non-existent. But such is not the case. The only reality perceived and asserted is the fact of non-existence in a case where a certain entity may be imagined or supposed to be. Combining our conception of this entity, *considered only as to its nature*, with the thought of non-existence, we exercise belief in connection with this combination. There is no incongruity in so doing. For we do not think of a thing as both existing and not existing at the same time; we simply displace from the positive conception of a thing the elementary thought of existence, and replace this by the thought of non-existence.

The distinction between positive and negative conceptions shows how we may exercise belief in connection with notions as well as in connection with propositions; because belief is possible whenever our thought in any way contains the element either of existence or of non-existence. The forming and holding of conceptions as setting forth fact or truth is what logicians have had in mind in teaching that "simple apprehension" is one of the three logical operations of the intellect. Whether we know something as a reality, or assume it to be such for the sake of argument, this apprehending and holding of a thing as true differs from the mere conceiving of a thing. It is even more than the conceiving of it as existing: it involves a real or affected belief in connection with our conception.

4. The division of notions, with reference to their fitness to correspond with realities, into the positive and the negative, is not an exhaustive division. There is a third class of conceptions, — *the formal, or schematic.* For should we, in conceiving of any entity, think neither of its existence nor of its non-existence, but only of its nature or characteristics, we might express this by saying that we think of it merely as a form, or schema. According to this use of language a "form" includes everything in an entity except its existence. This mode of conception is difficult of deliberate realization; but it occurs spontaneously sometimes, and especially whenever, after being ignorant about a thing, we learn whether it exists or not. For then, in our assertion respecting fact, we unite the thought of existence, or that of non-existence, with the schematic notion of the entity in question, and exercise belief in connection with this combination.

In every pair of conceptions contrasted with each other as positive and negative there is a part common to both; that part, when thought separately, is the formal, or schematic, conception. This mode of intellectual action has greatly escaped attention; it should have a place in every system of logic.

5. Another division of notions, less searching in its thought than the foregoing, distinguishes between the complete and the supplementary. A complete notion is one sufficient of itself to serve as a term — that is, as either subject or predicate — in a proposition; but a supplementary notion can only help to constitute a term. In the sentence "The white flakes of snow are falling gently on the grass," the adjective, the participle, and the nouns express complete notions, while the articles, the prepositions, and the adverb express supplementary notions.

Words significant of complete conceptions were called by the old logicians "categorematic," from the Greek κατηγόρημα, which signifies an assertion; while words whose force is merely supplementary were styled syncategorematic. A term which contains only one complete notion or categorematic word is said to be simple, but when several complete notions are combined in one term, it is called complex. In the above illustration both terms, namely, "the white flakes of snow" and "falling gently on the grass," are complex.

The distinction between complete and supplementary conceptions, and between categorematic and syncategorematic words, arises rather from our mode of employing ideas than from the essential nature of our thought; for direct and attentive thinking can give an independence to any conception whatever, and fit it for categorematic use. But this distinction prepares us to determine at once whether a proposition be fully formed or not, and what its terms may be. A thorough analysis of the component thoughts out of which terms or complete notions are constructed belongs to metaphysical psychology. Commonly in logic when we speak of conceptions we refer to complete conceptions.

6. This is especially the case in that division which distinguishes between *subjective and predicative* notions; for only a complete notion can be either subject or predicate.

Ordinarily, in making an assertion, we think of one thing, or entity, as existing, and then present another thing, either as existing, or as not existing, in some relation to the first. In saying "the snow is white," "the snow is not yellow," we think of snow as existing, and then assert that the quality indicated by "white" exists in the snow, and that the quality indicated by "yellow" does not exist in it. The first entity thought of in the assertion is called the subject, and the second the predicate; which terms are also applied to the corresponding conceptions. In common language, the subject is that about which some assertion is made, while the predicate shows what is asserted about it. Obviously, the meaning which logic thus attaches to the term "subject," is very different from that belonging to it in psychology, and according to which it signifies a thinking and sentient spirit.

The terms "subject" and "predicate" are applied, not only to things thought of in assertions, and to our conceptions of those things, but also to the words expressive of the conceptions. In the sentence, "The rose is red," the words "rose" and "red" are subject and predicate. But whatever may be the immediate application of these terms, they always refer to that use which we make of our conceptions when we affirm or deny one thing of another.

Two things which can be thought of as subject and predicate, and so as related to the faculty of judgment, may also be thought of *simply as related to each other*, and without reference to our assertion about them. In that light they have been named *substance and accident*, these designations being thus employed in a very peculiar way.

In logic, any entity whatever of which we conceive independently and about which we can make assertions — that is, anything whatever, as existing in predicable relations — is called a substance. In metaphysics we say that there are two kinds of substance, spirit and matter; in logic, spaces and times, powers and actions, changes, qualities, and relations are substances. When we speak of "the height of the column," "the beauty of the picture," "the wisdom of the judge," the height, the beauty, and the wisdom are logical substances, no less than the column, the picture, and the judge; for they may be subjects of predication.

In some discussions a distinctive name for the logical substance would prove advantageous; therefore we may occasionally speak of it as a *substantum*.

The term "accident," also, has a different meaning in this connection from what it has elsewhere, even in logic. For it is applicable to *any predicate entity whatever as united in being to a subject entity*. According to this sense the necessary properties of a thing, and even its essential attributes, are accidents. It would be well if some other word could be found to express this very general idea. Possibly the term "ascript," or "*ascriptum*," would serve the purpose. Then, when thinking objectively, the logician might speak of "substanta" and "ascripta"; though more frequently, and because he constantly considers the relation of things to thought, he will speak of "subjects" and "predicates."

CHAPTER V.

GENERAL AND INDIVIDUAL NOTIONS.

1. General notions. 2. The process of generalization. 3. The expression of general conceptions. 4. Realism, Nominalism, and Conceptualism. "Universals." 5. Individual, or numerical, difference. Specific difference. 6. Identity, numerical and specific. 7. The "principium individuationis." 8. "Individual" notions include (*a*) the singular, (*b*) the definite, (*c*) the indefinite, (*d*) the class notion; and are either unital or plural. 9. "All," distributively and collectively. 10. A restricted application of the term "individual."

1. An important logical distinction divides notions into the individual and the general. A notion is general when it is applicable to any of a class of similars simply on account of their similarity, and when it does not include the thought either of one object or of more than one. In saying "man is mortal" we do not conceive either of one man or of more than one, but only think that general notion, "man," which is applicable either to one man or to many. Should we say, "a man is mortal," or "any man is mortal," the words "a man" or "any man" would express, not a general, but an indefinite individual notion; which, however, is closely allied to the general.

2. Every general conception originates in a process called "generalization"; and this may be described as consisting of two steps, or stages. First, by an act of abstract thinking, we consider a number of objects so far as they are alike, withdrawing our thought from those respects in which they are unlike. This act is often preceded by a comparison of the objects, that is, by that process in which things are contemplated together for the purpose of perceiving their points of similarity and dissimilarity. This comparison is not always needed, and is easily distinguished from that act of abstraction in which the work of generalization properly begins.

The second step is the more essential one. In it, taking one or more of the objects as a sample or samples of the class of similars, we *drop from our conception all thought of individual difference* — all thought of number, whether of one or more than one; the conception which remains is a general notion. Thus, having perceived the similarity between many pieces of gold, we easily think of those many pieces under one plural conception, or we consider one piece as a sample of all; then, rejecting the element of individuality, we think and speak of "gold" in the general.

Some say that, in generalization, we conceive of "the many as the one" and of "the similar as the same." This language is incorrect and misleading. In generalization we do not regard a number of different things as if they were one and the same, but we discard all reference either to diversity and similarity or to unity and plurality, *and then think that one thought which remains.*

3. General notions, conceived independently, are expressed by common nouns, either without any addition or with the definite article prefixed. We say either "man," "gold," "wisdom," or "the pulpit," "the press," "the theatre." This use of the article indicates that the conception belongs to a class of objects well-known, and perhaps known in contrast with other classes somewhat resembling it, and, in so doing, it makes an addition to the general notion. For instance, "the pulpit," "the press," and "the theatre" are general designations applicable to well-known agencies of instruction, which also may be compared with one another. Since it is always possible to conceive, in this distinctive way, of things in the general, a choice is given between the simpler and the more precise form of expression. Some languages prefer the one; others the other.

The above-mentioned modes of conveying general notions by the use of nouns are the direct and proper methods. Other ways are employed, of which we shall speak presently, and which may be characterized as indirect and improper.

4. In using general thought and language we seem to be speaking about things, and we say that we are speaking about

things. This fact is the chief foundation for a doctrine, once very prevalent, that there are real entities corresponding to general notions as such. These entities were called "universals," and were considered eternal patterns, which, in some way, prepared for, and contributed to, the existence of individual entities. Thus it was held that man and tree and life and death and virtue and vice are universals, and that each of these imparts its nature to a large class of individuals as they come into being. The advocates of this doctrine were styled Realists, because they asserted the reality of general objects; they were opposed by the Nominalists, who taught that there are no such things even as general conceptions, and that universality belongs only to those names, or words, which may be applied to all the members of a class. A third doctrine, avoiding the extremes both of Nominalism and of Realism, has been called Conceptualism, because, while denying the reality of universals, it maintains that mankind constantly form and use general ideas. These ideas are not in their own nature general entities, but individual mental states. They are styled general because they are applicable to every member of a genus, or kind; for which reason they are also sometimes spoken of as universal notions.

The prevalence of Realism in former times and its influence, even at the present day, have been greatly promoted by the preference of man's mind for positive thinking and belief; we are naturally prone to believe that there are objects corresponding to our conceptions. This tendency favors Idealism, or the theory that the objects of the imagination really exist, as well as Realism. Language, too, falls in with both these delusions; for the very same words sometimes express actualistic conviction and refer to real objects, and sometimes express merely modes of thinking — imaginative or rational. Moreover, the fact that general conceptions and language are being continually *applied* to existing individuals with little notice on our part of any change in the method of our thought lends further aid to Realism. For the validity, or truthfulness, of general statements lies wholly in their applicability.

5. Let us now turn to individual conceptions. These are

distinguished from general conceptions because they are always modified by the thought of number, whereas a general notion excludes the qualification either of oneness or of plurality. An individual notion, such as "a dollar" or "dollars," always stands for what is, or may be, in strict literality, one thing or a number of things.

Every such entity is called an individual because it does not admit of "logical division." The general notion "dollar," as representing a class of things, may be divided into "gold dollar," "silver dollar," and "paper dollar"; and, in like manner, every genus may be divided into its species. But an individual dollar cannot be separated, even in thought, into a number of dollars, or things having the same general nature with itself. When, in the descending process of division, we come to the individual, we can go no farther.

The thought of individuality, like those of existence, non-existence, and entity, is simple and incapable of analytical definition. It is nearly identical with arithmetical "oneness" or "unity"; though oneness, in addition to individuality, includes the characteristic of quantity, and so sets forth every individual as a distinguishable quantum of entity.

When, along with a first one, another unit presents itself, we immediately perceive the relation of "otherness" existing between them, and so, considering them as quanta, we say that there are *two* individuals. All conceptions of number start from this beginning; hence the relation of otherness has been named *numerical* difference. Then, by a natural metonymy, that characteristic in every entity which is the basis of this otherness, is also called "difference." In other words, individuality, as the ground of otherness, is styled "numerical difference." So every individual may be said not only to be numerically different from every other, but also to have numerical difference in itself.

This difference is easily distinguished from that which exists between objects as being unlike each other. The latter is often called "diversity"; and it is also styled "specific difference," because it is the ground of dividing entities into species, or kinds. Two rain-drops might be so absolutely alike that they would differ only numerically; but there is specific as well as numerical difference between a rain-drop and a pebble.

6. Individual, or numerical, identity is that absence, or non-existence, of numerical difference which is perceived when any entity thought of once is compared with itself thought of again; it is a necessary attribute of every individual entity; it is what we mean by "sameness" in the strictest sense of the word.

Specific identity, on the other hand, is merely the perfect similarity which exists between two or more entities so far as they are members of the same species or genus; it is the "sameness" mentioned when we say that a thing may be done twice in the *same* way, or that all quadrupeds have the *same* bodily structure.

7. In scholastic times there was great discussion as to the "principium individuationis," or origin of individuality. The Realists held that individuals result from the conjunction of "universal" forms with the otherwise "undifferentiated matter" of being. But such forms and such matter are merely philosophical imaginations. The truth is that everything which exists has both individuality and definiteness in every part of its nature; these attributes begin and cease to exist as necessary elements of the entity itself.

8. Notions are styled individual because of the individuality of the things corresponding to them, and this equally whether a notion represents one thing or more than one. Hence in common language we might say that individual notions may be either singular or plural, but in logic we must say that this class of conceptions may be either unital or plural. For the term "singular," as we shall soon see, has a signification in logic quite different from that given to it in grammar, and therefore ought not to be used in logic in its grammatical sense. Such expressions as "a man," "men," "some men," "any man," "all men," "this man," "that man," "these men," "those men," "the man," "the men," "George," "the Georges," "President Cleveland," "his predecessors," "the presidents of the United States," represent individuals, and therefore set forth individual notions.

But while all unital and all plural conceptions are individual, and, under this title, are contrasted with general notions, the term "individual" is also employed sometimes in a more restricted application, as will be better understood after we consider four kinds, or classes, into which individual notions may be divided.

The first of these comprises those conceptions which logicians characterize as *singular,* and in which we conceive of an object as having marks peculiar to it, or of more than one object as having marks peculiar to them severally. For such ideas are unique, or singular, in their composition. These thoughts are often expressed by proper names, as when we speak of Niagara, the St. Lawrence, Washington, Cæsar, London, Paris; but they are also indicated by the common noun with the definite article or a demonstrative pronoun, it being then understood that the objects are known by means of marks peculiar to each of them, and not merely by some general character. If, in conversation respecting given persons or places, one should say, "I admire that man greatly," "I hope to visit those cities," the words, "that man," "those cities," would represent singular conceptions.

This same mode of speech, however, would express another class of notions if the objects mentioned were conceived of as definitely related individuals of a certain kind, yet without thought of peculiarities belonging to each of them. One might speak of "the President who was lately inaugurated," or of "the lawyer who has the case in charge," thinking of each only in his character as president or as lawyer, and conveying only this conception to others. Such ideas, because presenting objects as singularly related, though not as having peculiar natures, might be regarded as imperfectly singular. But they have been called "*definite*" individual notions, and, under this name, have been distinguished from singular notions; that is, from those perfectly and internally singularized.

A third species of notions to be mentioned here are the indefinite individual, or, more simply, the *indefinite.* For we may form a thought of a member of a class, or of more members than one, without determining our conception to any particular member or members. Such notions are indicated by the indefinite article and by the adjectives "any," "some," "several," "many," and other expressions of like meaning.

These conceptions, in themselves, are only the result of an indeterminate kind of thinking; but they are often used as substitutes for general conceptions. For example, the statement "a man — or any man — is mortal" may replace "man is mortal," because what is true of any man, taken at random, may be said of man in general. In like manner, the statement, "Some men live to a great age," may serve instead of "Man may live to a great age," because the probability, or contingency, in regard to man in general arises from the fact known indeterminately regarding some.

The fourth and last kind of individual conception is the *class notion ;* and this, like the others, may be either unital or plural. The unital class notion is indicated by the adjective "every"; the plural, by "all." The word "every" emphasizes the individuality of the things mentioned; the word "all," the universality of the statement about them: thus only we distinguish "every man must die" from "all men must die." In each case both individuality and universality are included in our thought.

9. For this reason it is important to notice a use of the adjective "all," which does not present the members of a class in their independent individuality, and therefore does not express a class notion. In saying "All the soldiers are brave men," we employ the word "all" *distributively*, as the logicians say, and consider the soldiers in their independent individuality. But should we say, "All the soldiers are the king's army," we would use "all" *collectively*, and would consider the soldiers, not merely as so many individuals, but as being united together; for it is only as united that they are an army. The Latin language provides for these two senses of the adjective by the terms "omnes" and "cuncti," this last being a contraction of "conjuncti." Whenever the subject of a proposition is a class notion, it must always be understood distributively, because a class considered collectively is no longer, for the purposes of logic, a class, but only an individ-

ual resulting from the union of individuals. "All men," as the family of Adam, or as the human race, are an individual, just as a congregation, a crew, a library, or a vocabulary is an individual.

The class notion is often used instead of the general notion when we wish to assert something as necessarily, and therefore universally, true respecting things of a given nature. When we say, "Every man is fallible," "All men are mortal," we give the form of individuality to the general truths that "man may be deceived," and that "man must die." The individualized assertion is an immediate consequence of the general truth, and has the advantage of being more closely related to actuality.

10. Having defined the four kinds of individual notions, we can now explain, in few words, that restricted application in which the term "individual" is sometimes used. It is that which contrasts the individual with the singular, and which therefore includes under the individual only the definite, the indefinite, and the class notions. For in all these we think of objects simply as individuals possessing a common nature. In this restricted sense individual conceptions are opposed to both singular and general conceptions.

CHAPTER VI.

PREDICATIVE NOTIONS.

"THE TEN CATEGORIES."

1. Subjective notions set forth *substanta*. Are of primary and of secondary conception. 2. Improperly distinguished as "concrete and abstract." 3. The ten categories of predication. 4. Substance as predicate. 5. Quantity. 6. Quality. 7. Relation. 8. Place, time, posture, condition, action, passion. 9. The substantialization of ascripts.

THE chief logical significance of conceptions arises from the employment of them as the subjects and predicates of propositions, but especially from their use as predicates. This involves many important modifications of thought.

1. All subjective notions set forth a "substantum," or logical substance; and their nature as substantal conceptions will be sufficiently illustrated if we divide them into those of primary and those of secondary conception; or, more simply, into the primary and the secondary. For while everything, of whatever kind, may be conceived of as a logical substance and as a subject of predication, some forms of entity are thought of in this way at once, while others are first conceived of ascriptionally, or predicationally, and only afterwards are treated as substanta. For instance, we think primarily of bodies and spirits — that is, of substances in the metaphysical sense — as substanta, and of the powers inherent in those substances as qualities to be predicated of them. Hence we say, "The scholar is wise," "The horse is strong." In like manner we conceive of a space as a substantum and of its size as an ascriptum, and say, "The room is large." Often, however, after some form of entity has been conceived of in the ascriptional way, we are led to think of it independently, and find ourselves doing so even while retaining in our minds a refer-

ence to our primary mode of thinking. In this way subjective notions of secondary conception arise. Thus from the predicates, or ascripta, in the cases given above, we may form the substantal notions "wisdom," "strength," and "largeness," or "magnitude."

2. The foregoing distinction is commonly expressed by the division of nouns, or of substantal notions, into the "concrete" and the "abstract." But these terms, though they indicate a difference, throw little light upon its nature. For the so-called "concrete" notion, if it be a *general* one — as "man," "animal," "matter," "spirit" — is formed by abstraction; and the so-called "abstract" notion, if it be complicated, involves a synthesis, or concretion, of ideas. For example, by synthesis we conceive of guiltiness as "a liability to penalty because of an infraction of moral law." Therefore, in a very natural sense, substantal notions of primary conception may be abstract, and those of secondary conception, concrete. This infelicity, arising from a conventional application of terms, illustrates a difficulty, which cannot always be avoided, in the expression of philosophical truth.

3. We now turn to the discussion of predicative notions. The earliest classification of these is one given by Aristotle. He says, "The Categories are ten in number, what a thing is (οὐσία), quantity (πόσον), quality (ποῖον), relation (πρὸς τί), where (ποῦ), when (πότε), position (κεῖσθαι), possession (ἔχειν), passion (πάσχειν), action (ποιεῖν)." The term κατηγορία originally meant an assertion, but here signifies a generic class, or *summum genus*, of things assertible. For, as Aristotle says, "Every proposition sets forth either 'what a thing is' or some other category."

This enumeration of predicative notions cannot be rejected as incorrect, yet is not closely connected with the laws of conviction. It belongs to a primary stage of logical theory, and is chiefly valuable as bringing before us, for further consideration, every form of ascriptional thought.

4. The first category, "what a thing is," was also named by Aristotle οὐσία, which term the scholastics translated by "substantia," or substance. The teachings of logicians regarding

this substance are confusing in the extreme, but we will arrive at its true nature if we remember that it is the form of predication expressed by a noun — that is, by the "noun substantive," as this was formerly distinguished by grammarians from the "noun adjective." In saying, "The man is a merchant," "Honesty is a virtue," the substance "merchant" is predicated of man, and the substance "virtue" of honesty. But while the term "substance" here clearly means a substantum, or logical substance, we cannot but observe that the application of it to the predicate of a proposition is accompanied by a modification of meaning. The subject of a proposition must always be conceived of independently before we can rightly say anything about it; therefore whatever is fit to be the subject of an assertion is a substantum in the full sense of the word. But no such independence of conception belongs to any predicative thought. The first of the ten categories may appear to have it, because this category originates in substantal conception, and is expressed by a noun. But *a noun used predicatively is preceded by a mental addition which destroys the independence of its conception.* For it then sets forth the predicate-substantum either as identical, or as not identical, with the subject-substantum; and this is quite a different thing from setting forth a substantum simply.

When we say that "the man is a thief," or that "the man is not a thief," we assert that the man is, or that he is not, identical (numerically) with a thief; we do not say merely that the man exists, and that the thief exists. Locke, and Leibnitz after him, perceived this mental addition, and hence, in their writings, the category of substance gives place to that of "identity and difference." There is, however, some advantage in retaining the old name. For the work of this form of predication is not completed in the assertion of identity or difference. Were that so, the category of "substance" would be only a specific form of the category of "relation." The true end of the predication of substance is to convey the information that a subject, already known as having one nature or aspect, has, or has not, another also. The statement, "The man is a thief," asserts that a subject known as a human being has the character of a thief; the corresponding negative statement denies that he has that character. In short, numerical identity or difference is here used to set forth the existence, or the non-existence, of a nature, or character, as belonging to a subject. Aristotle indicates this when he says that the first category shows "what a thing is," and in the name οὐσία; for οὐσία primarily signifies nature, or essence.

The secondary application of the term "substance" or "substantum," which we have now considered, gives rise to a secondary use of the corresponding term "ascript" or "ascriptum." Strictly and primarily every category of predication is an ascriptum, and, under this name, is contrasted with the substantum, or subject, to which it belongs. But when, in the classification and discussion of predications, we find one category called "substance," we naturally restrict the term "ascript" to the remaining categories, and thereupon we divide predicate notions into two comprehensive classes, the substantal and the ascriptional. Such language is scarcely avoidable when one may be speaking concerning the different kinds of predication, but it need not produce confusion, if we exercise care.

5. The second category — quantity — is used in asserting that something exists in a given degree or amount. In saying, "The road is ten miles long; the house is one hundred years old," we ascribe a definite age to the house and a definite length to the road; referring in each case to an appropriate unit of measure. And even in saying, "The house is old; the road is long," there is a tacit comparison with some standard. It is only this *measured* quantity that calls for a specific category. Quantity, simply as quantity, belongs to and characterizes every form of entity. It might be regarded as a kind of universal quality. As it may always be assumed, the predication of quantity, simply as quantity, seldom takes place.

6. The category of quality sets forth whatever does or may permanently mark an entity, and so be the ground of its classification with other entities similarly marked. This category is primarily expressed by the adjective, as when we say, "The man is wise; the table is round; the business is urgent." Ordinarily and properly the characterizing entity is attached

to the subject permanently; yet this condition may be dispensed with, provided only the mark be permanently connected with the subject in our conception. A dethroned king may still be thought of as a royal person; the general who has concluded a war successfully may be regarded for life as a victorious commander. In fact, the category of quality, like that of substance, is all-embracing in its power to use material; for any mode or combination of entity may be so used as to characterize the subject to which it is related.

7. The fourth category assumes that there are two or more entities, and then simply asserts (or denies) the existence of a relation between them. Thus setting forth the relation of cause and effect, we say, indifferently, "Fire is the cause of heat," or, "Heat is the effect of fire." The linguistic form of these statements belongs to the category of substance, yet they do not predicate substance, because their aim is simply to assert relation, and not nature, or kind. We may also express relation by saying, "Fire produces heat, or is productive of heat," provided our intention is not to assert that heat is being produced, or that fire can produce it, but the fact that heat is produced by fire. Relations are primarily expressed by prepositions, but are often set forth in this secondary way by nouns, or verbs, or adjectives.

In speaking of relations as existing *between* entities, our language is based on the circumstance that the conception of a relation comes intermediately between those of the relata. In strict truth, however, a relation is not an intermediate entity, but is composed of two parts, or relationships, one of which resides in each of the things related. This doubleness, or plurality, appears in the relation of husband and wife, of agent and instrument, of cause and effect, of equals, of unequals, of the container and the contained, and in all other relations.

8. The next category is that of place. Some have objected to this category that it is merely a specific mode of the category of relation. But it is, or at least may be, more than this. "The king lives in a marble palace," sets forth both that there is a marble palace and that the king lives in it. In like man-

ner, a relation and something more are expressed by the category of "time." "The marriage took place last Thursday," indicates both that an event occurred at a certain date, and that a certain time has elapsed since that date.

The categories of "position" and "possession" might better be named "posture" and "condition." They also have a doubleness. To say, "John sits," or, "John is resolved," sets forth a posture of body or of mind in which the parts of the body or the thoughts of the mind are adjusted to each other, and are, moreover, externally related. For one sits on some seat, and is resolved on some conduct. In the same way the sentence, "John is well, and John is wealthy," indicates first the existence of health and wealth, and then the condition in which John finds himself as the possessor of these blessings.

Finally, the categories of "action" and "passion" both set forth the operation of some power, but the one in relation to the agent or instrument, the other in relation to the thing acted upon. Therefore these, also, are duplex.

9. Having familiarized ourselves with the natural forms of predicative thought, as presented in the "ten categories," and having seen that predicative conceptions may be divided into two general classes, the substantal and the ascriptional, we must not fail to note, in conclusion, an important point. This is that either the substantal or the ascriptional mode of predicative conception may take the place of the other. Especially we must understand how a statement with an ascriptional predicate may, by a slight addition, be changed into an equivalent statement with a substantal predicate; for a change of this kind often takes place necessarily in the course of our reasonings. When we say, "Some men are wise; therefore some wise beings are men," this reasoning is valid only because we replace the ascriptional proposition, "Some men are wise," by the substantal proposition, "Some men are wise beings." So, in the syllogism "Man is rational; every rational being is accountable; therefore man is an accountable being," the argument would not be conclusive if it were not lawful to replace the ascriptional term, "rational," by the substantal

term, "rational being." Moreover, in the final proposition of this syllogism we have found ourselves at liberty to adopt the substantal form of assertion, though the ascriptional form might have been retained.

The thought of the "being," or "entity," which is added in these modifications of conceptions is that of the substantum to which the ascript belongs. We have the right to make this addition, because, when any subject has an ascriptional predicate, it may, of course, be identified with itself as a substantum having that ascript, and, when it has not a given ascript, we can say that it is not a substantum which has it.

This process might be called the *substantialization of ascripts*.

The reverse process, of de-substantialization, consists in dropping the thought of substance. Instead of saying, " Man is a mortal," we say, "Man is mortal." This change occurs frequently, but is of less logical consequence than the other.

CHAPTER VII.

PREDICATIVE NOTIONS.

"THE FIVE PREDICABLES."

1. Defined and enumerated. 2. Genus and species here signify natures, not classes. 3. Species, essence, definition, and nature, distinguished. 4. Difference, as a predicate, is not the relation, but the ground of it. 5. Property. Generic and specific. Often becomes attribute. 6. Accident. Here opposed to essence and property, not to substance or subject or being. Separable as regards the nature; separable or inseparable as regards the object. 7. The "predicables" are used only when logical connection is conceived exactly. 8. Attributes. Adjuncts. Qualities.

1. A SECOND division of predicative notions given by Aristotle is known as "the five predicables." This classifies all the possible predicates of any subject, not with reference to their own differences, as in the categories, but *according to their exact connection with the nature of the subject.* The distinctions thus presented are quite important; because the force of a proposition, either as setting forth truth or as a premise in argument, varies with the mode in which the predicate is logically related to the subject, or, as Aristotle would say, with the mode of the inherency of the predicate in the subject.

Logicians formerly taught that every predication used in reasoning not only conforms to one of the ten categories, but also to one of the five predicables — in other words, that it not only asserts substance, quantity, quality, relation, or something else, of a subject, but also presents the predicate employed as related to the nature of the subject in one or other of five ways. They expressed this by saying that every proposition sets forth either the genus, the species, the difference, the property, or the accident, of a thing; and they held that all reasoning arises in connection with these last-mentioned modes of apprehension; which, therefore, by way of pre-eminence, were called the predicables. These views are extreme. Predications, and reasonings by means of them, may take place without reference to any predicable. But it is true that these modifications of assertive thought are often employed in our more thorough thinkings, and that they have an important function in the apprehension and statement of truth.

2. The first predicable is *genus* (γένος). This term frequently signifies a class of similars in which other classes of similars, differing from one another, are comprehended. According to this sense the genus, "forest-tree," comprehends oaks, beeches, maples, elms, and so on. In the present connection genus means, not the generic class, but that nature which belongs to every member of it. When we say that the oak is a forest-tree, and think of it as having the nature of all forest-trees, and distinguish this nature from the peculiarities of the oak, we predicate genus of it.

Since it is part, though not all, of our conception of the nature of the oak that it is a forest-tree, the predication of genus does not, in this case, add to our **knowledge of what an oak is, but only makes a part of** our knowledge explicit. If, however, we were ignorant concerning the nature of an oak, or of anything else, the predication of genus would enlarge our information, and would not be merely explicative of a conception already entertained. The predication of "species" or of "difference" may, also, be employed in either of these ways.

The question may be asked, "Is the nature asserted in the predication of genus individual or general?" We reply that it is either, according to the character of the subject. The predicate of a general subject is necessarily general, and that of an individual subject individual. Should we speak in general and say, "The oak is a deciduous tree," all our thought would be general; a similar assertion made about this or that oak would be individualized throughout. We do, indeed, say that the individual tree has a generic nature, but this use of language is secondary and metonymical. It does not mean that the tree has literally a generic nature, or a general nature of any description, but only that part of the individual nature corresponds to a generic conception.

The second predicable is designated "species" (εἶδος). This term often signifies a subordinate class of similars, but, in the present connection, it means the nature which characterizes such a class. We say that, while man is, according to genus, an animal, he is, according to species, the rational animal. Thus it appears that "species" comprehends "genus" together with a "difference," by which the given species is distinguished from other species of the same general kind.

3. The predication of species, however, is not the mere assertion that a subject has a certain distinctive nature united with a generic nature. It implies that *the nature predicated (the species) is the whole nature which the subject has in common with other entities, so far as that nature may be conceived of by us.* It would not give the species of horse to say that the horse is a quadrupedal mammal. This would only present a genus, though it would be a subordinate genus formed by the union of the higher nature "mammal" with the peculiarities of the specific nature "quadruped." To give the species, we must add those particulars regarding form and motion, parts and uses, which complete the conception "horse," as entertained by us.

It is not, indeed, necessary for the purposes of definite and conclusive thinking that we should give all the particulars that enter into this conception. Very often one or two or three of the distinguishing features are sufficient, as representatives of all the specific peculiarities; nevertheless it remains true that the predication of species is the predication of the whole nature conceived of.

This was taught by Aristotle when he said that to give the species is to give the *definition* of a thing; and it is involved in the doctrine that "species" and "essence" are identical, or nearly so. The predication of species and that of essence present exactly the same truth; they differ only in that the former directs the mind to the substantal form of conception, while the latter dwells on the attributal, or qualitative. To say that man is the rational animal (giving this predicate its full substantal force) asserts species; to say that man is rational and animal, or that he has a rational and an animal nature, gives his essence. Either form of statement, however, may be said to set forth either essence or species.

Let us note here that, in logic, the term "nature" is closely related to "essence," and has precisely the same meaning whenever we refer to the whole nature, or constitution, of a subject. But we may speak of a generic nature, as when we say that the oak in its generic nature is a forest-tree; while we do not speak of a generic essence. A nature, therefore, may be only part of an essence.

4. The third predicable, "difference," has been designated also "specific difference," and is thus opposed to "numerical difference." Its office is to present, not the relation of difference between different species, but the foundation of this relation; namely, that peculiarity, or collection of peculiarities, which belongs to a species and distinguishes it from others in the same genus. Among plane figures bounded by curved lines the "difference" of a circle is that every part of the circumference is equally distant from a point within; and the "difference" of man among animals is "rationality."

Aristotle says that genus and difference are interchangeable and identical. In a certain sense this is true. We may think first of the genus "animal," and then of the difference "rational," which distinguishes one species of animal; or we may think first of the genus "rational," and then of the difference "animal," which distinguishes one kind of rational beings. In like manner, a circle and a sphere have the generic character that in each every part of the boundary is equally distant from a point within, the differences of these figures being that the one is solid and the other plane; but, were circle compared with square, or sphere with cube, in either case, the genus given above would become difference and the difference genus.

Nevertheless, though what is now genus may become difference, and what is now difference may become genus, genus and difference are not the same; nor is the predication of the one the predication of the other. A nature is genus as the foundation of resemblance between species; it is difference as the foundation of diversity; so that the same nature cannot be both genus and difference in reference to the same two specific classes. Genus and difference as such are not interchangeable with each other, but they may exist together for the same reason that two men, who are related as creditor and debtor, may, in their relations severally to two other men, be debtor and creditor.

The "specific difference," now under consideration, is easily distinguished from that "individual difference" which belongs to entities simply as such, and whether they differ in nature or not. Moreover, when two individuals, being of different kinds, are said each to have its specific difference, this difference is a part of the individual and is itself an individual thing. Yet it is not, on this account, what we call "individual" (or numerical) difference. To assert this latter is merely to say that one thing is not another; but to assert the former is to say that one thing is unlike another.

5. The fourth predicable is named "property," this term being thus used in a strict and technical sense. A property is that which is not included in an essence, or species, but which yet is necessarily, and therefore universally, connected with it. Thus it is the property of man to be a religious being, and of a plane triangle to have its three angles equal to two right angles.

Property being inseparable from essence, our conception of an essence may easily be enlarged by incorporating with it that of some property; upon which addition property ceases to be property and becomes attribute — that is, an essential characteristic. For this reason, and because our conceptions frequently vary in comprehensiveness, it may sometimes be difficult to say whether some necessary ascript be a property or an attribute. For example, since every quadrilateral figure is quadrangular too, one might ask, "Is it a part of the essence, or only a property, of such a figure to have four angles?" The answer is that this is either a property or an attribute, according to the manner of our conception. Mostly, for the sake of simplicity, the mind selects just so many leading and permanent marks as are sufficient to distinguish a class of beings from all others, and excludes all remaining ascripts from its idea of essence. This is especially the case in the forming of definitions. Yet it is not invariably so. In conceiving of a triangle we think of three angles as well as of three sides, and recognize the angles as entering into the essence of a triangle, though they are inseparably involved with the sides. The only way to determine whether a necessary characteristic be a property, is to ascertain whether it be something additional to our conception of the object. Accordingly, we say that it is the property of a circle to contain a greater extent of surface for the length of its boundary than any other plane figure, and of man to be a member of political society.

While property is always attached to an essence, or species, this connection may immediately relate either to that generic part which the species has in common with other species, or to some peculiarity in the "difference" of that one species. Hence properties are of two sorts, the generic and the specific, or differential. Mortality is a generic property of man, as an animal; the power of using language is a specific property of man, as the rational animal.

6. The fifth, and last, predicable is "accident." It is that which pertains to an object, or entity, yet which is not necessarily connected with the nature of the object. The faculty of language and the power of laughter are properties of man, while the act of laughing and that of speaking are accidents; because a man is not always laughing and speaking.

Moreover, we must rank with accidents any ascript concerning which we cannot tell whether or not it is necessarily involved with the nature, or essence, of the subject; although such an ascript is not an accident in the full and proper sense of the term. For it resembles accident, and it is unlike property and attribute, in this important respect, that it cannot be inferred from the mere existence of the subject. But, in the full sense of the word, that only is an accident which is known to be separable from a nature, or essence.

This separability, however, means only that an accident is not a necessary consequent or concomitant of the nature of the subject. It does not mean that every accident is inseparable from the object which may have the given nature. With respect to this object an accident may be either separable or inseparable. It was accidental to Voltaire, considered as a man or as a genius, to be born in France; yet this fact was inseparable from the man. It was, also, an inseparable accident of Socrates, as the father of Grecian philosophy, to be a statuary. So, on the supposition that there are no human beings except those born on this planet, it would be an inseparable accident of man to be a native of the earth; for, so far as their nature is concerned, human beings might be born elsewhere. The inseparable ascript of a class of things, however, is seldom conceived of as an accident. It is found to be connected in some way with the nature of the subject, and is regarded as a property.

We must not leave this fifth predicable without noting how the term "accident," as here opposed to "genus," "difference," "species," and "property," is much more limited in application than when it is opposed to "substance," or "being." The accidents of a thing simply as an entity include everything whatever that can be predicated of it. The reason for this is, that an entity, simply as such, is not necessarily one kind of thing rather than another, so that every addition to our thought of it is, in a sense, accidental. This wide signification of "accident" — as equivalent to "ascript" — is easily distinguished from its ordinary logical meaning, though the two are by no means disconnected.

In discussing the five predicables we have used objective rather than subjective language, following the ancient manner of speaking. We have mentioned genus, species, difference, property, and accident, rather than generic, specific, differential, proprietal, and accidental, conceptions. In the primary and literal sense of words it is not things, but notions, as representative of things, that are predicable. Yet the ancient mode of expression serves to remind us that the logician always considers thought objectively, even while he may be studying our varying conceptions of the same thing or kind of thing. For these vary only because we contemplate an object now in one aspect and in one set of connections, and now in another aspect and in another set of connections.

7. The classification of predicative conceptions under the five predicables applies only to those cases in which the exact logical connection of the predicate with the nature of the subject is part of our thought. We can, and often do, make assertions without determining whether the predicate be a genus, or a species, or a difference, or a property, or an accident. The doctrine of the predicables, therefore, is not so widely applicable as that of the categories.

Every "predicable" presupposes one of the categories, and then makes an addition to it. For it presents some ascript, not simply, but as related in some one of five ways to the nature of the subject. The predicables, therefore, are of the same radical nature with the category of relation; yet they are properly discussed by themselves on account of their functional connection with all the categories, and because of their logical importance.

Evidently the end and use of these complex modes of conception is to state the manner in which any ascript is logically related to the nature of a subject; for they always set forth something, either as the whole essence of a substantum, or as included in the essence, or as attached to it. The predicables are those forms of predicative thought which we naturally employ after obtaining thorough information regarding the logical relations of a subject; while the categories are those more simple and primary forms of conception by which truth, whether individual or general, is set forth without reference to the logical connection of things.

8. Logicians speak only of five predicables; yet some other conceptions — especially "attribute" and "adjunct" — are of the same general character. Anything included in either genus or difference — that is, any part of the species, or essence — is, technically speaking, an attribute. Attributes, therefore, are either generic or differential. A nature consists of the sum of its attributes. Whatever is connected with a nature

without being a part of it, is an adjunct. Every adjunct, of course, is either a property or an accident.

"Quality" is also a kind of predicable, and nearly the same as attribute. It is properly that mode of conception which sets forth "what kind of a thing" the subject may be; that is, which assigns to the subject a generic (or a differential) as distinguished from a specific, nature. But, with a somewhat wider use of language, *whatever does or may permanently characterize is called a quality.* Hence properties sometimes receive this name, because by enlarging our conception they may be taken within the nature. When quality is used to set forth nature or character simply, and without reference to logical connections and classifications, it is a "category," not a "predicable."

We have seen that, for most logical purposes, the categories may be reduced to two classes, — the substantal and the ascriptional, — and that these may be made to replace one another in assertions. In accordance with this, we now add that any one of the predicables may be expressed by either substance or ascript. We may say either, "John is a man," or "is human"; "is rational," or "is a rational being"; "is a biped," or "has two legs"; "is a European," or "was born in Europe"; "wrote that note," or "was the writer of that note," intending by our language to set forth either genus, difference, property, or accident. We more naturally express genus and species substantally, and the other predicables ascriptionally; but either may be expressed either way.

CHAPTER VIII.

THE DEFINITION OF NOTIONS.

1. Clearness and distinctness radically the same. Definition and division defined. 2. Definitions are either essential or accidental. 3. Some, necessarily, are accidental, or relational. 4. Essential definitions are either exhaustive or selective; 5. Scholastic or notational; 6. Adequate or inadequate. 7. Nominal and real definitions. 8. The essence of a thing is either (*a*) its whole form, or constitution, (*b*) its form *so far as conceived of* by us, or (*c*) the prominent and *important part* of its constitution. 9. "Substantial forms." Singular essences.

1. CLEARNESS and distinctness should not be contrasted as radically different. Distinctness is simply clearness considered as enabling us to make correct distinctions. Whenever a thing is clear, it is also therein distinct. Sometimes, however, we say that an object is apprehended clearly when its several parts and boundaries are perceived; and that it is apprehended distinctly when these are exactly and perfectly perceived. In this contrast distinctness is the highest attainable degree of clearness.

Definition and division are processes whose chief aim is to render our conceptions as clear, or distinct, as possible. Each contemplates every object as a whole; but definition regards in turn the several elements of an object, as they are severally related, and then presents these in a connected statement; while division studies a number of objects more or less similar in their nature, with respect to their points of agreement and of difference, and then arranges them according to their agreements and differences.

In saying that we attentively consider objects in these operations, we mean only that we scrutinize things in *idea*, not that any entities are actually examined. If, indeed, we have no reliable knowledge of an existing object or set of objects, we should make it the subject of our enquiries and observations.

The science of logic directs one to such investigations, and gives useful directions concerning them. But the processes of which we now speak presuppose a certain kind or degree of knowledge, and merely aim to give the ideas we have so definite a shape, that we may both be free ourselves from confusion or unreadiness, and be able to guide others into a correct understanding of our conceptions. The attempt to form a definition or division may bring to light the inadequacy of our information, and so lead us to remedy that deficiency. But definitions and divisions of themselves pertain properly only to the perfecting of notions formally. Finding that one knows, or conceives, of an actual or possible class of things, and without enquiring whether the things really exist or not, definition gives an internal, and division an external, distinctness to our conceptions.

The final result, or product, of either process takes the same name with the process itself, and is called a definition, or a division. In each case, also, it is expressed by an identifying statement. A definition identifies an object considered concretely, or without analysis, with itself as distinctively characterized, while a division identifies a generic class with the species contained in it. The character, however, of both processes is fairly stated when we say that a definition sets forth the nature, or essence, of a thing, while divisions set forth the different species of things that are of the same genus.

2. Let us now confine our discussion to definitions, and let us consider four distinctions which may illustrate their nature and use. First, we say that *definitions are either essential or accidental*. The essential definition sets forth the nature of a thing directly, and is the result of an analysis. It either enumerates attributes or gives genus and difference. The accidental definition, on the other hand, sets forth the nature of a thing indirectly and by means of a suggestion. It makes use of properties or accidents. Should we say that "man is the rational animal," this would be an essential definition; it would be an accidental definition to say that "man is the religious," or "the political," or "the talking, animal." In mineralogy the diamond would be defined by its essence as a

brilliant stone formed by the crystallization of carbon, but by one of its accidents, if it should be called the most precious of gems.

The question whether a definition be essential or accidental cannot be determined simply from the definition itself, but is connected with the aspect under which the object may be made the subject of our enquiries and assertions. Ordinarily an essential conception embraces what appear, from the most general point of view, to be the permanent and distinctive characteristics of a thing. But sometimes our study is confined to a particular point of view, so that our discussions relate only to some limited aspect of the subjects considered. Especially a science or art may describe objects in a technical and peculiar way, while yet such descriptions must be accepted as essential definitions, so far as that branch of knowledge is concerned. For they set forth the natures of things according to that science. Linnæus, thinking of man only as an animal, defined him as "the two-handed mammal." In chemistry laudanum is a vegetable extract of a given molecular constitution; in Materia Medica it is a poison operating in a specific way. Each science may determine its own distinctive conception of an object; and then the essence is that in the object which corresponds to this conception.

The term "accidental," as applied to definitions, is used in a wide sense which relates to all adjuncts, whether properties or accidents. For properties, even more than accidents, are used in accidental definitions. Indeed, these might be advantageously styled *relational, or adjunctional, definitions*, as they use what is related, or joined, to the essence.

3. Some logicians, with some reason, say that the accidental definition does not define, but only determines, a conception. The question is one of terms. Ordinarily and pre-eminently, a definition is an analytical statement; yet, if every proposition which fixes the meaning of a word by giving, directly or indirectly, the essence of a thing, may be called a definition, then we must admit accidental definitions. These are especially necessary in discussions concerning things simple and incapable of analysis. For in no case can the duty of giving

a clear conception be avoided by saying that the thing is simple and ultimate, and does not admit of definition. Every conceivable entity can be determined by means of its adjuncts or relations; and, if it be simple, it ought to be defined in this way. Hence it is proper to say, "Space is that abiding kind of entity in which all other things exist, and in which motion takes place"; "Time is that fleeting kind of entity during which events transpire, and by reason of which they are related as past, present, and future"; "Belief is an intellectual state differing from conception, specially conditioned on the thoughts of existence and non-existence, and expressed by the assertive proposition"; "Sensation is a psychical experience caused by the action of certain nerves, and is the condition of our perception of material things and qualities."

Accidental definitions are also useful when the object, though complex, is not easily described, or when there is no need for specific description. One might say that the guillotine is the instrument by which capital punishment is inflicted in France. But since this statement would not give the "essence" to one unacquainted with the construction of the guillotine, it does not deserve the name definition as well as those statements by which simple natures are indicated relationally.

4. The next distinction pertains to essential definitions only, and, in the remainder of this discussion, these will chiefly occupy our attention.

The essential definition is either *exhaustive* or *selective*. In the former case it sets forth all the attributes belonging to a subject as having a specific nature; in the latter, only the more distinctive and important characteristics. It would be an exhaustive definition of a circle to say that it is a plane figure bounded by a curved line which returns into itself, and which is everywhere equally distant from a point within. Promptitude might be exhaustively defined as the habitual disposition which leads one to decide and act at once when the proper occasion has come. But it is a selective definition to say that man is the "rational animal," for we always think of man as having a certain bodily shape and size, and as having a practical and affectional, as well as a rational, nature. Defining coal as "a black combustible mineral widely used as fuel," this would sufficiently express our ordinary conception of that substance. The same would be true if we should say that coal is "a geological vegetable deposit largely composed of carbon."

In one sense substances, especially material substances, admit only of the selective definition, for we always regard them as possibly having other attributes than those which we know them to have. Iron has chemical, medicinal, and magnetic powers of which the ancients knew nothing, and probably has other qualities not yet discovered. Yet all these qualities are allowed for in our conception of the permanent nature of this metal. Our notions of substances, therefore, though they may be accurate and reliable, are never analytically complete. But our definitions of powers, spaces, times, figures, relations, and other non-substantial entities may be exhaustive; that is, they may give every part of the constitution of the object as conceived of by us.

5. In the next place the essential definition may be either *scholastic or notational*. The scholastic definition, which is that commonly discussed in logic, is effected by giving the genus and difference of a thing. In other words, it uses two comprehensive conceptions, one of which presents the essential attributes of a genus, and the other the distinguishing attributes of a species. This mode of definition is naturally followed when the generic characteristics of a new kind of thing are easily known, while some care is necessary to determine its difference from other species of the same class. We perceive at once that an oak or a beech, a pine or a cedar, is a tree; then we proceed to say how it is distinguishable from other trees. We at once recognize oxygen or hydrogen as a gas; then we enquire what are its differential peculiarities. Moreover, in lectures and discussions, definition by genus and difference is a clear and compact mode of statement.

At the same time this is not the only proper mode of definition. The notational method, which simply enumerates attributes (or essential marks) without any reference to genus and difference, is equally correct with the scholastic, and is often employed to advantage. It is less artificial than the scholastic.

6. Another distinction discriminates between *adequate and inadequate* definitions. A definition is adequate when it imparts a distinct understanding of the nature to be defined. Exhaustive definitions are adequate when they are expressed simply and without superfluous additions. Selective definitions suffice when they present the most important and distinctive attributes. Hence the scholastic definition gives both genus and difference; for neither of these by itself would distinguish, and each is supposed to contain some fundamental marks. The two together present the whole nature; though they may not give it exhaustively, but representatively. Hence, too, insignificant marks cannot be the basis of definition; for they are not "essential" in the sense of being important. "Man is the two-handed mammal," is an inadequate definition of human beings, unless we limit our thought to the sphere of natural history. The question of the adequacy of a definition depends on the question whether it clearly sets forth that conception of a thing which we wish to use.

7. Finally, definitions are either *nominal* or *real*. Logicians differ in their explanations of this distinction. Some say that the nominal definition sets forth the meaning of words, and the real, the nature of things. But every definition, though pertaining immediately to notions, necessarily explains also terms and natures. Others teach that the real definition sets forth more of the nature of the thing than is implied in the name, while the nominal deals with a less comprehensive conception. The difficulty with this explanation is that the signification of a term expands when our conception of a thing expands; so that there is no good ground to distinguish our more contracted conceptions as nominal.

The true difference between nominal and real definitions seems to be that the former *simply explicate notions, without teaching that objects really exist of the nature described,* while the real definition *implies that there are objects corresponding to it.* Should we describe a dragon as a winged serpent breathing flame, or a mermaid as an inhabitant of the sea, half woman and half fish, these definitions would be merely nominal. Indeed, every definition must be treated as nominal until we know that it sets forth a real nature. Thus some political law or institution, being merely supposed to exist, might be defined in order that its probable operation might be accurately discussed.

Most definitions, however, and especially those used in science, are real, because they are intended to be applicable to actually existing objects. Therefore, also, they are more than mere definitions; they are assertive, as well as explicative, propositions; and on this account may be made the grounds of actualistic inference. Having learned that saltpetre is nitrate of potash, we can say that any piece of saltpetre has this composition and all the properties flowing from it. At the commencement of discussions respecting matters of fact one should see to it that the definitions laid down are not merely clear explanations of conceptions, but also truthful representations of things.

8. This discussion concerning the definition may be concluded by an enquiry which may render more exact our understanding of the objective significance of this form of thought. Let us consider *what is meant by "the essence" of a thing*. For any statement which distinctly gives the essence, either directly or indirectly, is therein a definition.

The word "essence" is a Latin term, said to have been invented by Cicero in translation of οὐσία. It is more restricted and definite in its use than the Greek word; yet it admits several varieties of meaning. The doctrine of the essence may be presented in explaining these varieties. First of all, we might say that the essence of a thing is its entire being, or entity, considered analytically, or as constituted of related parts, or elements. This would follow from the common statement that "the essence is that which makes a thing to be what it is," provided we take these words in their full force. For no entity would be what it is if any of its elements were wanting. In this sense the essence of a thing is identical with its entire form, or constitution. For *any entity can be viewed either indeterminately, that is, simply as a thing, and without distinction of parts or elements; or its parts may be individually and definitely conceived of.* Entity as viewable in the one way has been called "*matter*," and as viewable in the other way has been called "*form*"; while, as conceivable in both ways at once, it is both matter and form. Essence, therefore, may sometimes signify the entire form, or constitution, of a thing. This, however, is an extreme use of the term.

Commonly by "essence" we do not mean the whole constitution, or make-up, of a thing, but only *that constitution so far as it is actually conceived of by us*. For, ordinarily, our thought of an object is not exhaustive of everything contained in the object, but takes in only such characteristics as have engaged our attention. Hence only so much of the constitution, or constituents, of an entity, as we conceive of determinately is called the essence; while the rest is treated as so much matter, or indeterminate entity. For example, all the particulars included in our conception of water would be, for us, the full essence of water; though we may allow that other attributes may belong to the nature, or form, of water, in its absolute totality.

Finally, essence may mean something less even than the nature of a thing so far as conceived of. For often a few of the prominent and controlling elements of that nature are taken as the representatives of all, inasmuch as the rest may be inferred or supposed where these are found; in short, we limit our thought to what might be called the representative, or symbolic, essence. Hence man is defined as the "rational animal."

The first of the three significations now given may be rejected as improper; it is better expressed by speaking of the entire nature, or form, of a thing. The other two significations are often employed. Using them only we distinguish the *complete* and the *representative* essence; the former being set forth by the exhaustive, and the latter by the selective, definition. For an exhaustive definition gives every part of a thing as conceived of by us, though not necessarily every part of a thing.

According to the foregoing the idea of essence is properly

of more restricted application than that of form, or nature, or constitution. But, of course, when we speak or think of an essential form, or nature, we mean simply an essence.

9. Ancient logicians used to mention the "substantial form," and they distinguished this from the substance (or substantum) on the one hand, and, on the other, from the essence, and even from the entire form. So far as the phrase embodies truth a substantial form is a substantum considered as having a given essence or form, that is, as more or less analytically, or determinately, conceived of. But substance, as contrasted with substantial form, was entity viewed independently, yet not determinately, but only as ready for determinations — that is, entity viewed merely as entity, merely as a "thing." This distinction has been found needless in modern philosophy. By substance, or substantum, we now mean anything whatever considered independently and as fitted to be the subject of predication, whether it be already characterized — that is, conceived of definitely — or not.

Some have taught that the essence of a singular thing does not include any of its singular characteristics, but only such as may belong to it as the member of a species; and that, therefore, the species of an individual thing and its essence are always identical. It is true that, in our discussions, things are commonly conceived of as having specific natures, and are defined only in that light, and as having a specific essence. We might say, for example, of some assertion that it is, essentially, a lie. Yet — so far as we can see — any statement in which the leading features of a singular conception are set forth may be said to give the essence of the singular object, and may be called the definition of that object. Thus it would define Bucephalus to say that he was "the spirited horse which Alexander the Great tamed and rode."

Here, too, a false distinction concerning essences may be briefly mentioned. It pertains only to material or spiritual essences, and not to those of substanta, or logical substances, in general. Locke, in particular, teaches that every substance has two essences, the "*nominal*," which is perceivable by us and is set forth in names and definitions, and the "*real*," which is the underlying but incognizable basis of the knowable essence. This doctrine is connected with the mistaken view that substance is the incognizable substratum in which active and passive qualities inhere. Both theories must be rejected as philosophical fictions. Every substance may have qualities as yet unknown to us, but we have no reason to say that these qualities are unknowable, or that they are the origin of those known. The truth seems to be that we may understand the real and ultimate nature of any kind of substance, though, it may be, not exhaustively. There may be a difference, of course, between the constitution of a substance so far as known and that total constitution which is only knowable. But this does not justify the theory of a knowable and an unknowable essence.

CHAPTER IX.

LOGICAL DIVISION.

1. Aims to render conceptions distinct and definite. A succession of synthetic acts. 2. To be distinguished from didactive and from rhetorical division. 3. Indispensable that the dividing members exclude one another. 4. Division should relate to one "principle." 5. Specially useful if the principle be an "added mark." 6. Certain exceptions to this rule. 7. The dividing members should be co-ordinate; but this is not an absolute law. 8. The division in certain cases should be exhaustive. 9. Dichotomy, or "infinitation," a comparatively profitless division.

1. The second logical process by which our conceptions are rendered distinct is called "the division of notions." This process takes any collection of conceptions which have a common generic part and co-ordinates them — or rather the *concepta* corresponding to them — into subordinate genera and species, according to their agreements and differences. After "paper" has been defined as a "fibrous material manufactured into sheets from the pulp of rags, straw, or wood, by a process of spreading, pressing, and drying," one's conception, either of paper in general or of any kind of paper, becomes more distinct, if we enumerate the various kinds, and compare them with one another. Reflection on the peculiarities of writing-paper, wrapping-paper, wall-paper, building-paper, hard-pressed paper, drawing-paper, and tissue-paper, as well as on the character common to all these varieties, frees one's ideas both of paper and of kinds of paper from an indefiniteness which often beclouds unelaborated thought.

The name "logical division" literally indicates the mental separation of a generic class into its component species, but it is metonymically used to indicate the formation of specific conceptions from a generic conception by the successive addition of differences. For we speak of the division of a generic notion. This process, however, though involved in the division of a class, is not really divisive, or separative; it is additive, or synthetic.

2. Logical division is not to be confounded with that didactive division by which the heads of a treatise are distinguished and arranged. These processes are of a kindred nature, but they aim at different results. Logical division sets forth methodically the principal agreements and differences which exist within some generic class. With this end in view, we may regard that class in various lights, and divide it in different ways. A people may be classified according to their diversities in political opinion, in religious belief, in business occupation, in sex, or in age, or in color, or in any other respect; and, in order to a thorough knowledge of that people, they should be considered in all these respects. But in planning an account of them one need not follow any one of these divisions, but should simply arrange his thoughts in the order best suited for the conveyance of his information.

At the same time it is to be borne in mind that a scientific treatise gains in lucidity if its more fundamental distinctions be expressed by logical divisions; and commonly the order of its discussions wisely refers to one or more of these divisions. For instance, President Woolsey, in one of the opening chapters of an excellent treatise, divides International Law first into Public and Private; then again, into the Law of Rights and Obligations and the Law of Claims and Duties; and then, finally, into the Law in Time of Peace and the Law in Time of War; after which he adopts an order of discussion based on this last division. But he might have chosen some other order had he seen fit.

The oratorical arrangement of thought, which aims at conviction and persuasion rather than instruction, is yet more separated from logical division than any arrangement which is merely didactive.

3. In order that logical division may fully effect its purpose, several rules are given, of which, however, one — and only one — is so fundamental as to admit of no exception. This is, that *the dividing members of the genus must be exclusive of one another.* In other words, the specific natures in any logical

division must be such that no two of them can belong to the same individual subject. To divide the genus "book" into the "folio," the "quarto," the "French book," the "German book," the "reader," the "dictionary," and so forth, would violate this rule; because the same book might have two or more of these natures. Again, to divide "moral actions" into "right" and "wrong," or into "public" and "private," would be a good division. In either case conflictive natures would be presented. But it would not be a division to say that "moral actions are either right or wrong or public or private"; for these four species of moral actions, though distinct, are not separate.

It is useful to compare individuals of different species together when the species are not conflictive; or to compare the same individual as being of one species with itself as being of another species also. A quarto and a dictionary might be compared in a way to bring into prominence their common part and their peculiarities. But this is a more analytical and delicate operation than the one under consideration. Logical division recognizes that natural separation of classes which results from the fact that mutually repugnant natures may be successively united to a generic nature; and which, therefore, immediately shows, with respect to the members of each species, both what a thing is and what it is not. This process, which at once characterizes and distinguishes things, is easily apprehended; and is of great service especially at the beginning of a discussion.

4. A second rule, nearly equal in importance to the first, is that *the division should refer to one principle, or fundamentum*.

The word "principle," here, is used in a special sense. Often it signifies a general truth used as a premise in argument, or reasoning; here it designates a generic characteristic to which a specific difference may be immediately attached. The common character "sex" belongs to all quadrupeds; and, with respect to sex, they may be divided into male and female. Mankind, with reference to that capability of culture which distinguishes them all from brutes, are divided into the enlightened, the civilized, the semi-civilized, the barbarous, and the savage. In saying that a division should rest on one principle, or basis, we mean that the same characteristic should be used successively with the several differences, so as to form the several species. The same thing is meant when we say that division should rest on one *fundamentum*. But this application of the word "fundamentum" must be distinguished from its use in connection with comparison, and with the perception of relations generally. In the latter case a fundamentum is one of the things related, considered as the foundation of the relation; so that comparison calls for two fundamenta at least; logical division requires but one.

The rule that only one principle of division should be used is somewhat auxiliary to the rule that the species should be mutually exclusive. For example, if we take only one fundamentum for the division of the human family, — say race, or country, or language, or religion, or sex, or age, or condition in life, — it is easy to form a classification in which the members will be exclusive of one another. But if a division uses first one and then another principle, the species will be likely to overlap. To divide the people of the British Isles into English, Scotch, Welsh, Irish, Protestants, and Roman Catholics would be a violation of the first rule resulting from a neglect of the second.

The principal advantage, however, of adhering to one fundamentum is not the aid which this rule gives to others, but its direct effect in adding to the clearness of our conceptions. This result takes place in a twofold way. First, the consideration and comparison of the different species *brings the principle of division into distinct view*. Dividing men into Jews, Mohammedans, Christians, and Pagans, we see that man in general is a religious animal, and that this religiousness is something different from any particular form of faith. Then, secondly, through comparison of the difference which belongs to each specific religion with the peculiarities of the other forms, we are led to perceive exactly the character of each. Divisions thus made often have the effect of definitions, and may even be considered a kind of "accidental" definition. For they present both the principle of the division and the constituent species in determinative relations to one another.

5. Some say that the "fundamentum divisionis" should always be an essential mark — an attribute — of the genus to be divided; others say that it should be an "added" mark, that is, some adjunct of the genus. The truth is that it may be either; though more frequently it is an added mark. In the above illustration the religiousness of man is a property, not an attribute. Classifications based on an added mark are especially helpful when we would study some general nature in an enlarged aspect which includes more than its essence, or ordinary definition; as, for instance, when we consider human beings as religious; for in this we have added something to their essence.

But that an attribute as well as an "added mark" may be the fundamentum of a division may be easily shown. When "rectilineal figures" are divided into "three-sided," "four-sided," "five-sided," and so on, the differences are successively attached to the attribute of having straight sides. After the same manner we divide color into white, black, red, blue, yellow, and so forth. In this latter case, since the nature of the genus may be regarded as simple and as containing only one attribute, it is all used as a "fundamentum divisionis."

6. We must add, however, that certain classifications of natural objects seem to attach their differences to a variety of fundamenta. These are cases in which Nature herself in diverse ways has made additions to a complex of attributes. We divide the genus "gas" into oxygen, the *life-supporting* gas; hydrogen, the *lightest* of gases; nitrogen, the *most inert* gas; and chlorine, *the colored* gas. Vertebrate animals are divided into mammals, birds, fishes, and reptiles. Quadrupeds — though they may be divided according to one principle, as, for example, into graminivorous, carnivorous, omnivorous, and so forth — are ordinarily classified as the elephant, the rhinoceros, the hippopotamus, the horse, the cow, the dog, the cat, and so on. In such divisions the leading peculiarities of each species are attached to the genus not immediately, but through a fundamentum used only, or chiefly, for that species. For instance, all vertebrate animals bring forth young; mammals bring forth their young alive: all have some natural covering and

style of locomotion; birds are covered with feathers and fly in the air: all have a proper habitat where they live and breathe; fishes live and breathe in the water: all have a bodily structure which determines their modes of activity; reptiles are so made that they crawl upon the belly. A division, therefore, which uses different fundamenta in the formation of its species cannot be rejected as incorrect, provided only its members exclude one another; as they do in the above classification of animals.

7. A third rule of logical division is that *the dividing members should be co-ordinate with one another;* in other words, the several species into which any genus is immediately separated, should show the *same amount of difference* added to the common character. This direction is especially to be observed in natural history and in all classifications which are designed to set forth the nature of things as exactly as possible. It may, however, be often dispensed with in divisions which aim only at clearness of statement. It would be a violation of this rule if animals were divided into invertebrates, reptiles, fishes, birds, quadrupeds, quadrumanes, and the biped, man. The first of these species would not have enough added difference, and the last three would have too much, to make them co-ordinate with the other classes mentioned, and with each other. Correct division first distinguishes animals into the vertebrate and the invertebrate; then vertebrate animals into reptiles, fishes, birds, and mammals; and then mammals into quadrupeds, quadrumanes, and bipeds.

But frequently distinctness of statement does not require that the component kinds should be co-ordinate; the attempt to make them so may even savor of undue refinement. Plane triangles are rightly divided into the equilateral, the isosceles, and the scalene, though, according to the rule, we should first distinguish triangles which have some sides equal from those which have no sides equal, and then subdivide the first class into the isosceles and the equilateral. In like manner, grammarians properly classify words as monosyllables, dissyllables, trisyllables, and polysyllables. Nothing would be gained by dividing them, first into those of one syllable and those of more than one, and then subdividing this latter class.

8. One other rule remains: *the division should be exhaustive;* in other words, the dividing species, taken together, should equal the whole genus divided. But this completeness of classification, though an excellence to be desired, is indispensable only in certain cases.

Sometimes an exhaustive enumeration *is needed for practical purposes.* A treatise on mortgage investments should indicate all the various ways in which a loan might prove unsatisfactory. The money might be lent on security of insufficient value; or on land with a bad or doubtful title; or on property burdened with previous mortgages; or where taxes, court-judgments, or mechanics' liens detract from the security. The papers might not describe the property adequately; or they might not be legally drawn-up and executed; or they might not have been recorded duly and in proper time. Then, also, faithless or incapable agents, tricky or shiftless borrowers, unwise local laws, or inefficient tribunals, may render it undesirable to make a loan. A person with money to lend should think of all these things, and so avoid, as far as may be, every cause of loss or annoyance.

Complete enumeration is also necessary *when we would reason disjunctively to a definite conclusion.* In a trial for murder the prosecutor might say that one man may kill another either accidentally, or in self-defence, or through passion, or through deliberate malice, and then argue that it could not be in either of the first three ways, and therefore must be deliberate murder. This reasoning would be defective because of incomplete enumeration. The killing might have resulted from insanity, or from some hallucination.

In science exhaustive classifications give satisfaction because of the completeness of view and treatment for which they prepare. Yet such classifications can be hoped for only in those departments of knowledge whose boundaries and provinces have been accurately ascertained. Many fields of investigation — such as botany, mineralogy, natural history, and others, — do not admit of it. In these sciences we classify all we know, but expect to find species as yet unknown.

9. A truly exhaustive division is based on a pervading and

comprehensive intelligence, every species in a genus being distinctively conceived of. We do not, therefore, give this character to a process sometimes called "dichotomy"; because it divides either all things in general, or all the individuals of a genus, into two classes : it is sometimes also called "infinitation"; because it conceives of one of the classes as "infinite," or rather as "indefinite," in extent and character. Evidently all things whatever are either of some given kind or things not of that kind; and any genus consists of the individuals of some one species and others not of that species. Man and not-man may be said to make up the universe, and animals are either quadrupeds or not-quadrupeds. This dichotomy is often useful in argumentative statements, but it throws little light on the nature of things, and is of small value as a logical division. It may also be used as a test when the question arises whether each member of a division excludes all the rest.

CHAPTER X.

PROPOSITIONS AND PREDICATIONS.

1. Judgment involves both conception and conviction. The proposition defined. 2. A proposition is not always an assertion. Predication defined. 3. Propositions are either enunciative or assertive; 4. Affirmative or negative; 5. Presentential or inherential (the predication proper); 6. Verbal or mental; 7. "Categorical" or "conditional"; 8. Actualistic or hypothetical.

1. HAVING discussed the notion, or conception, we pass to the judgment, or assertion. This is the second general mode of rational mental action. Ordinarily judgment signifies the formation of a probable belief: in logic *it is the act of forming any conviction whatever*. The most absolute demonstrative conclusion, and even the immediate perception of fact, are included under this term. Should one look upon a rose and say, "The rose is red," this would be a judgment.

Both the primary powers of the mind, thought and belief, are exercised in judging. We first conceive of a thing existentially,—that is, think of it as existing or as non-existent,—and then, exercising conviction along with this conception, we assert that the thing is, or that it is not.

That form of words in which a judgment is fully expressed is called a proposition; and this name is also given to that construction of thought in which a judgment is embodied. Propositions in words, however, are important only as they are related to propositions within the mind.

A mental proposition is essentially of the same nature with an existential conception, but it is more analytical and brings the idea of existence, or of non-existence, into prominence. To believe in "God" is the same as to believe in "God as existing"; and this belief is fully expressed by the proposition, "there is a God," or "a God exists."

2. Propositions are not always assertions. In themselves

they are only the forms of thought which judgments employ. At the beginning of every criminal trial two propositions are presented to the jury, "the man is guilty" and "the man is not guilty." Neither of these embodies a judgment at first, but one or the other does so when, after the hearing of evidence, it becomes the verdict of the jury.

While the term "predication" also designates the form of thought employed in judgments, it implies further that the existential conception, or proposition, is not only entertained, but believed in, or asserted. The difference between "predication" and "proposition" may be traced to the primary signification of these terms. Originally a proposition signified a statement placed before one for his acceptance or consideration, while a predication meant a statement in which some fact was made known or published. Hence "proposition" came to refer specially to the thinking person, and "predication" to the objects thought of. But as a statement about things is more closely allied to the exercise of conviction than a statement of one's ideas, the notion of assertion was always retained in connection with the thought of predication, while it was often disconnected from the thought of a proposition.

3. We are now prepared for certain distinctions by which propositions are logically divided, and which illustrate both the radical nature of judgment and the various modes in which judgment takes place.

The first of these, emphasizing the truth that even existential thought may be unattended by conviction, divides propositions into *the enunciative and the assertive*. At the opening of every debate it is proper to state the "question." Then two contradictory propositions are before the "house"; for example, that "a protective tariff should be maintained," and that "it should not be maintained." Up to this point the propositions are mere enunciations; either of them becomes assertive when it is upheld by its advocates as true, or is accepted as true by those who have weighed the arguments.

The assertive proposition does not always set forth a thing as fact, or even as indisputably true, but it always expresses belief of some kind or degree. It is never the mere statement of a thought, or of a question. Should one venture the opinion that "$100,000 would be a comfortable fortune," this would be an assertive proposition, though it would not set forth fact, but merely what would be fact if one had that amount of money.

The science of logic deals mainly with assertions. The enunciative proposition is studied exclusively for the light which it throws upon the assertive.

4. A second division of propositions relates primarily to their assertive use, but is also applicable to them when merely preparations for assertion. It separates them into *the affirmative and the negative,* the former setting forth the existence, and the latter the non-existence, of things. Every predication is either affirmative or negative. That such is the case is evident; when we are in doubt concerning a thing we cannot make any assertion about it; but so soon as doubt is displaced by any conviction at all, we must think and believe either that the thing is so, or that it is not so.

Because assertion is the most important function of propositions, their fitness for assertion has been technically named their "quality"; and so it is taught that, in quality, every proposition is either affirmative or negative. We should note that this distinction applies to those propositions called "conditional," as well as to those ordinary propositions which make assertions in a direct way, and which are called "categorical." "If Gabriel be a man, he is mortal" is an affirmative, "If he be an angel, he is not mortal" is a negative, conditional proposition. The "quality" of such a statement appears, not in its antecedent, but in its consequent; for this is the assertive and essential part of it. But that remarkable form of the conditional proposition which is styled the "disjunctive," has the peculiarity of not being limited to one kind of "quality"; it has both kinds, though only because it is itself a complex of simple conditionals. "The number is either odd or even," means, "If the number be odd, it is not even; and if it be even, it is not odd; but if it be not odd, it is even; and if it be not even, it is odd." Two of these propositions are negative, and two are affirmative.

5. A third distinction between propositions, though easy to make, is not easily expressed in any terms yet employed by philosophers. This must excuse our use of barbarous language when we distinguish the *presentential* from the *inherential* proposition. The presentential proposition may be defined as a simple existential statement; it says simply that a thing exists or that it does not exist. Thus we may assert that space exists, or time, or that there is money or virtue; or the opposite of these things. But the inherential proposition sets forth a thing as *existing in some relation;* or as non-existent in that relation. When we say, "Virtue is praiseworthy; money is useful," we assert, not the existence of virtue or of money (that is assumed), but the existence of the qualities of praiseworthiness and usefulness as belonging to virtue and money. In like manner, to say, "The tree grows; the man thinks" sets forth the present existence of certain actions on the part of the man and of the tree; while, "The tree does not grow; the man does not think" sets forth the non-existence of those actions. To say, that "diamonds are stones" assumes the existence of both diamonds and stones, and then asserts that identity exists between these two classes of things to such an extent that every diamond is a stone; while to say, "Coal is not stone" denies — that is, asserts the non-existence of — identity.

By far the greater number of statements are inherential. These, too, are more important, logically, than the presentential; for all inference and reasoning arise from the perception of things as existing in connective relations; not from a knowledge of their mere existence. On this account, and because the forms of language make little or no distinction between the two kinds of statements, only inherential propositions have hitherto been recognized by logicians. Aristotle teaches that "*a proposition is a sentence in which one thing is affirmed or denied of another.*" This definition does not apply to the presentential proposition; because a sentence which simply asserts the existence or the non-existence of a thing, cannot be said to assert one thing of another. Most writers since Aristotle, merely adopting his words, say, "Judgment is that act of the mind whereby we affirm or deny one thing of another." Even those who have differed from him have failed to perceive the distinction between presentential and inherential statements. Locke taught — and very many have followed him — that judgment is *the perception of agreement and disagreement* (or of congruence and conflict) *between two ideas;* and others, in our own day, define judgment as *"the perception of relations."*

We prefer the ancient doctrine to either of these. Aristotle is right in saying that we judge respecting things, "affirming or denying one thing of another." Even when we judge about objects that are absent, or that are merely supposed, it expresses the truth better to say that we are judging about things than to say that we are comparing ideas: and, while every inherential proposition does set forth something as in relation to a subject, the point asserted in the great majority of such statements is, *not that the relation exists, but that the predicate-object exists as related.* When we say, "The man is wise; the man speaks," the points of assertion are, not the relations of the wisdom and of the speaking to the man, but the wisdom and the speaking themselves. Aristotle's teaching would have been satisfactory if he had explained that the affirming and the denying one thing of another are simply the assertion of the existence, and the assertion of the nonexistence, of the predicate-object in its relation to the subject-object. Yet, even with this explanatory addition, his doctrine takes no note of presentential assertions.

Although the word "predication" may originally have designated both presentential and inherential statements, usage and the objective reference of this word have rather confined its application to the inherential proposition. This is especially indicated by our use of the term "predicate"; which is the name we give things as logically "inherent." The term "predication," therefore, might be used exclusively for inherential assertions; or at least, we might say that these are predications-proper, and then style presentential statements improper predications.

6. A fourth distinction is valuable chiefly for the light

which it will throw on other distinctions yet to be considered; it discriminates between *verbal and mental propositions*. Aristotle, speaking of the theory of demonstration, says that this theory pertains not to the external word, but to the word within the spirit — οὐ πρὸς τὸν ἔξω λόγον, ἀλλὰ πρὸς τόν ἐν τῇ ψυχῇ λόγον. He also entitles his treatise concerning propositions περὶ ἑρμηνείας, or De Interpretatione, showing that, in his view, every proposition must be scrutinized, if we would obtain its true meaning.

It is, however, to be remarked that the "external word," of which Aristotle speaks, is something deeper and more internal than mere language. It is rather that thought which language immediately and naturally expresses, and which, therefore, may be called "verbal," in the absence of any better name. Frequently this thought is not the thought really intended to be conveyed, but is only indicative or suggestive of it; and nothing is of more importance than to distinguish between that form which thought may take immediately before it is expressed in language, and the very essence and intent of the thought itself. For no mistake could be greater than to suppose that a form of statement which originally and properly expresses one mode of assertion may not be used with another logical signification.

This may be illustrated from the fact that the grammatical subject and predicate of a sentence are often different from the true logical subject and predicate. In the sentence, "Cæsar conquered the Gauls," the grammatical subject is "Cæsar," and the grammatical predicate, "conquered the Gauls"; these would be the logical subject and predicate if the object of the speaker were to inform one who already knew of Cæsar, of Cæsar's conquest of the Gauls. But if it were only known that the Gauls had been conquered, and the question was, "Who conquered them?" then "Cæsar" would be the logical predicate, and the logical subject would be given in the grammatical predicate. In this case, if grammatical expression were conformed to logical meaning, we would say, "He who conquered the Gauls was Cæsar." If, again, the hearer knew of both Cæsar and the Gauls, but was ignorant of what the one did to the other, the logical form of the sentence would be, "What Cæsar did to the Gauls (subject), was to conquer them (predicate)." The logical subject is that which is conceived of as already known; the predicate is that which is asserted of the subject.

For further illustration let us take the proposition, "All men are mortal." Grammatically and verbally, this simply sets forth a fact respecting all existing men; just as if one should say, "All the men in that village are industrious": mentally and logically, it sets forth *the law,* "Omnibus est moriendum," that "*man, whenever and wherever he may exist, must die.*"

This use of verbal propositions to express the true and inner thought of the mind arises, partly from the imperfection of language and partly from the disposition and ability of men to make the same linguistic form serve several purposes. The circumstances of a statement generally suggest the right interpretation of it.

7. These remarks bring us to a fifth distinction; by which propositions are divided into *the categorical and the conditional.* With Aristotle a categorical was simply an affirmative proposition; but his disciple and successor, Theophrastus, gave the name to any statement, whether affirmative or negative, which is made unconditionally, or rather without any *expressed* condition. All subsequent writers follow Theophrastus in this use of language. They also oppose to the categorical the "conditional" proposition; the condition here referred to being not an actual, but a supposed, condition. By a conditional proposition, therefore, we are to understand *an assertion expressly depending on some hypothesis, or supposition, this* dependence being indicated by "if," *or some other word introducing the supposition.* "If iron is heated, it will expand; if a man is wounded, he may die," are conditional assertions. So also is the statement, "A whole number is either odd or even"; for this expresses conditions by "either" and "or."

"Iron is a metal; birds are oviparous; food is necessary to life; twice three are six; all men are mortal; some snakes are venomous; an equilateral triangle must be equiangular;

a straight line can be drawn from one point to another in a plane; France is a republic; Carnot is president," — these are categorical propositions.

But while logicians rightly distinguish categorical and conditional statements, many overlook the fact that *this division pertains to verbal, and not to mental, propositions.* The distinction is commonly presented as if it gave not merely the superficial form, but also the deeper nature of our thought. That a more profound discrimination is needed should be evident to those who know that conditional statements can be changed to a categorical form, and that categorical statements can often be turned into conditionals. This latter is always and especially the case when the categorical assertion sets forth a general truth.

Instead of saying, "If iron is heated, it will expand," we can say, "Heated iron expands"; while the sentence, "A wounded man may die," is equivalent to, "If a man is wounded, he may die." Even the disjunctive conditional can be expressed categorically; though not by one proposition, it being a complex of assertions. For example, the four conditionals which compose the assertion, "A number is either odd or even," may be replaced by the following categoricals: "A number not odd is even; a number not even is odd; an odd number is not even; and an even number is not odd."

Conversely, all the categorical examples given above, except the last two (concerning France and Carnot), may be changed into conditionals; thus, "If a mineral be iron, it is a metal; if an animal be a bird, it is oviparous; there must be food, if we would live; if one doubles three, he has six; if a being is human, he is mortal; if a man is a scholar, he may be wise; if the creature be a snake, it may be venomous; if there be an equilateral triangle, it must be equiangular; if two points be assumed in a plane, they can be connected by a straight line lying in that plane." Not only may many categoricals be thus changed into conditionals, but their true and internal nature is more fully set forth by the conditional than by their own form of expression. In short, while conditionals need no interpretation, categoricals are very often secondary modes of statement which are unconditional only in their verbal, and not in their mental, significance.

8. Let us now close with a distinction which applies the same thought to propositions, or assertions, in their inner nature, which the distinction of categorical and conditional applies to their superficial character. In the fifth place, we say that propositions are either *actualistic or hypothetical*. Here, as the word "conditional" is used to designate propositions hypothetical in form, we take the liberty to apply the term "hypothetical" to all propositions which are hypothetical in their true character, whether they be expressed in hypothetical language or not. The peculiarity of such predications is, that they do not assume the reality of their antecedents or subjects; and therefore also do not assert reality for their consequents or predicates. It is an hypothetical statement to say, "If a farmer be diligent, he will prosper"; and to assert, in the general, "The diligent farmer will prosper," is equally hypothetical. For this last, although referring to the fact that some farmers are diligent, is not intended to assert fact, but only to affirm that, if a certain antecedent should exist, a certain consequent will exist also. But to say regarding some known individual, "That honest and diligent farmer will prosper," would be actualistic. Propositions "conditional" in their verbal character are always hypothetical mentally, and therefore need no interpretation; but, as we have already seen, very many categorical statements are essentially hypothetical.

Actualistic propositions are those which express belief in fact; that is, in the actual existence, or non-existence, of a thing. They do not always assert reality simply, or purely, as when one might say, "God created the world"; they may set forth fact as necessary, or they may present something as probably, or possibly, a fact.

All presentential assertions are actualistic, but inherential propositions have this character only when they assume the real existence of their subjects and express some conviction, great or small, qualified or unqualified, regarding their predicates. "There is bread in the cupboard; there is no money in the house; there may be flour in the barrel; the judge is impartial; the witness may be mistaken; that prisoner cannot be guilty; he must be innocent," are actualistic assertions.

The verbal form, or superficial significance, in designation of which the name "categorical" is given to ordinary propositions, primarily pertains to actualistic conviction. That independent mention of the subject with which it begins naturally appears to present the subject as actually existent. This circumstance obscures the fact that the majority of categoricals are essentially hypothetical.

The true state of the case, also, is further obscured, because the categorical form of statement, even when used hypothetically, does not wholly lose its original force. In saying, "War is an evil," we *refer* to wars and their evils as realities even while our mental aim is not to set forth these things as facts, but to state the law — or general sequence — that if, or whenever, there may be a war, it is an evil. The actualistic implication in such a general statement is no essential part of it, but merely a concomitant of it.

In addition to these natural causes of mistake logicians, by a defective use of terms, have positively inculcated the error of denying the hypothetical character to every categorical statement. For they have confined the name "hypothetical" to "conditional" propositions; while that extensive class of assertions which are mentally hypothetical, but which may be either of the conditional or of the categorical form, has been left without designation, and almost without recognition. Let us remember that a proposition categorical in form may be hypothetical in its inner meaning.

CHAPTER XI.

CATEGORICAL PREDICATIONS.

1. Verbally free from hypothesis. 2. The predication-proper analyzed. The common, and the true, doctrine of the "copula." 3. Origin of the verb "to be" as copula. 4. Categorical assertions are simple or compound; 5. Substantal or ascriptional; 6. Affirmative or negative; 7. Universal or particular. "Distributed" and "undistributed" terms. 8. Indefinite, and singular, propositions. 9. Hamilton's quantification of the predicate. 10. Definitions, divisions, and "exclusive" predications. 11. The "pure," and the "modal," categoricals.

1. CONDITIONAL propositions are simply hypothetical statements fully expressed: they will be further considered hereafter. Categorical propositions are statements which are free from any expression of hypothesis: they are actualistic in form, though often hypothetical in fact. The principal varieties of categorical assertion constitute an important topic of study. Before discussing them, let us consider a part, or element, which is essential to every inherential proposition; and which logicians call the "copula."

2. Every predication-proper comprises, first, the subject about which the assertion is made; secondly, the predicate asserted of the subject; and thirdly, the copula, which is the verb "to be," expressed or understood, and agreeing grammatically with the subject. In saying, "The daffodil is a flower; the daffodil is yellow," the copula is expressed: in saying, "The daffodil blooms," it is understood, or rather it is united with the predicate and concealed in it. For this last sentence means, "The daffodil is blooming."

The name "copula" is connected with the view that the verb "to be" in predications-proper indicates, not the existence of anything, but the union in thought of the predicate with the subject; this conjunction of things being considered the essence of affirmation. Moreover, since the days of Aristotle, it has been held that the word "not," added to the copula so as to signify negation, indicates the mental separation of the predicate from the subject.

These teachings are erroneous except so far as they imply that the copula expresses the essential, or differential, part of assertion. Affirmation is not a uniting, nor is negation a separating, of things in thought. The true doctrine is that affirmation sets forth the existence, and negation the non-existence, of the predicate-object. In presentential assertion affirmation and negation take place simply upon the conception of the subject as existing and as non-existent, and without any joining or sundering of things.

As to the "conjunction and separation" of things in predications-proper, we have three remarks to make. *First*, composition, or synthesis, is not especially necessary for affirmation, but is the condition of any predication whatever. For a predicate must be conceived of as related to a subject before it can be either affirmed or denied with respect to that subject. In affirming, "The candle burns," an action, conceived of in an appropriate relation, is asserted to exist in that relation; in the statement, "The candle does not burn," precisely the same thought-combination of entities is used, though it is accompanied by the assertion of non-existence. *Secondly*, negation, even when completed, does not involve any separation of things in thought; though it may be naturally followed by the dismissal from our minds of the rejected predicate. To assert the non-existence of a thing which has been conceived of as in some relation is no more a mental sundering than the extinction of a flame is the removing of it from the candle. The flame is not taken away; it is extinguished where it is. So the predicate is not removed in thought, but is simply asserted not to exist. *Finally*, the assertion of separation, so far from being a negation, is only a specific case of affirmation. To exist in separation is incompatible with that kind of union of which we ordinarily conceive, and may be used to deny such a union; but it is itself expressed by a combination of the predicate with the subject; for this may be based on any relation whatever. Moreover, as it positively asserts separation, it implies the existence of both the things separated. To say, "The horse is free from disease," taken literally, implies that both horse and disease exist in separation from each other. But as the real point to be asserted is merely the non-existence of disease in the horse, and as this would follow if the disease were at a distance, the affirmation of the separation is used figuratively, and merely denies the existence of the disease. In such a case there is no real assertion of separation at all.

We repeat, therefore, that the function of the verb "to be," alone in affirmations, and accompanied by "not" in negations, is always to set forth existence and non-existence; and that, in predications-proper, this function is conditioned on the union of things in thought. The mental separation of things — the conceiving of them as apart — is only indirectly and accidentally connected with negation.

3. Here the question may be asked, How is it that the verb "to be" sets forth the existence of the predicate, when it agrees grammatically with the subject? This singular mode of speech must have originated in the days when men first began to make assertions. It seems to be a linguistic device in the use of which certain verbs, which were primarily used to indicate the existence of subjects, had predicate-facts attached to them by way of grammatical limitation. The different parts of the verb "to be" in our own language, and of corresponding verbs in other languages, are traceable to roots signifying to breathe, to live, to stand, to remain, to grow, to be born. In short, existence, originally, was expressed by words which signified more than mere existence. For example, men said, "The man breathes; the man lives," meaning only, "The man exists." Then it was found easy and natural to express predicate-facts by adding something to the same verbs which had been selected to indicate the existence of the subject. It was said, "The man breathes free; lives virtuous; remains at home; is born rational; has grown wise; has stood firm," the point of assertion being only, "The man is free; is virtuous; is at home"; and so on. In like manner, negation was expressed by giving each verb a negative limitation; as, "The man

breathes not free; lives not wise; remains not at home." In the course of time the connecting verbs, entirely losing their proper significance, indicated only the existence (and the non-existence) of the predicate.

In concluding our remarks concerning the copula, we must add, what is allowed by all, that the reference to time, expressed by the different tenses of the verb, is no part of the copula, but must be attached either to the subject or to the predicate, as the analysis of the sentence may require.

4. Let us now consider the leading varieties of categorical assertion, following, as in previous discussions, the method of logical division. First, categorical statements are either *simple or compound.* The simple categorical has only one subject and one predicate; the compound is a condensed expression for several simple assertions, which are independent of each other yet have a common subject or predicate. "Cæsar came and saw and conquered" is equivalent to "Cæsar came; Cæsar saw; and Cæsar conquered." "Cæsar, Charlemagne, and Napoleon were conquerors" means, "Cæsar was a conqueror; Charlemagne was a conqueror; and Napoleon was a conqueror." The rules of logic pertain immediately to the simple categorical, and therefore presuppose a resolution of compound propositions into their components.

5. In the second place, categorical assertions are either *substantal or ascriptional.* This distinction concerns only the verbal form of predicates. The subject of every proposition is a substantal notion; but a substantal assertion depends upon a substantal predicate. "Man is rational; the gentleman lives near me," are ascriptional propositions: "Man is a rational being; the gentleman is my neighbor," are substantal predications. Evidently this distinction pertains to superficial thought; because any ascriptional statement may be replaced by a substantal equivalent; and because this process may be reversed. We make such changes almost unconsciously in our ordinary reasonings. For instance, the substantialization of the predicate occurs whenever we "convert" an ascriptional proposition in order to show what it authorizes us to say regarding the predicate considered as the subject of an asser-

tion. For ascriptional statements are convertible only after being changed into the substantal. Therefore, replacing "No men are perfect," by "No men are perfect beings," we convert and say, "No perfect beings are men."

Because every ascriptional proposition may be given substantal form, and for some uses must assume this form, it has been taught that all propositions, when fully expressed, are substantal: in other words, that every predication is essentially the affirmation or denial of identity. This is an extreme opinion. It is not sustained by the analysis of mental facts.

6. With respect to "quality," or "assertivity," categorical predication is either *affirmative or negative*. This distinction, being connected with the essential nature of propositions, has been discussed already.

7. We proceed, therefore, to that distinction which pertains to the "quantity" of propositions and which separates them into *the universal and the particular;* for this division is important only as applied to categorical assertions. "Quantity" is the extent of the applicability of a proposition; and it is of two kinds, according as the subject-notion indicates a whole class of things or only part of a class. In the former case the subject is said to be "distributed," and the proposition is "universal"; in the latter case the subject is "undistributed," and the proposition is "particular." "All men are mortal," is an universal proposition; "Some men are wise," is a particular one.

The term "distributed" *might* designate the conception of *any* number of things conceived of as individuals, and not collectively. In saying, "All men are the family of Adam," the expression "All men" is used collectively; but in, "All men are mortal," it is used distributively. And if we should say, "One man could not lift a ton, but some — or several — men could," the word "some" would have a collective force; but in, "Some men are learned," its force is distributive. Therefore, in a broad sense, not only the word "all," but also the word "some," may be used distributively. In logic, however, a term is not called distributed unless it be distributed *fully*, or used in its widest possible application; and a term, though conceived distributively, is yet considered undistributed if it be applied to anything less than the whole class. Hence the rule that universal propositions distribute the subject, while particular propositions do not.

8. Under the head of quantity, besides the universal and the particular, Aristotle mentions *the indefinite proposition, and the singular*. These, however, should not be treated as separate species. Quantity is indefinite when it is either not determined in thought or not expressed in language: in the latter case it is also styled "indesignate." "Gold is metallic; horses are used for riding," are of indefinite quantity, but a little thought makes one of these universal and the other particular. The quantity of a predication can always be determined; it is universal if we know that the predicate belongs to the subject as distributed, and particular if we cannot say so. For the quantity of an assertion is based on what we are able to say.

A singular proposition is one with a singular subject. It makes an assertion about one object conceived of with its individual peculiarities, or about more than one object thus conceived of. For example, "The Gracchi were the head of the agrarian party." Logicians class the singular with the universal proposition; as its assertion pertains to every individual mentioned, whether one or more. This view may be accepted, if singular predications be allowed quantity at all. But the truth is that the "quantity" discussed in logic serves a special purpose as related to general, or generic, classes of things, but has no proper significance in connection with singular assertions.

The quantity of a proposition may be given either *a unital or a plural* expression. Instead of, "All men are mortal," we can say, "Every man," or "any man," or "man always," is mortal. Instead of, "Some serpents are venomous," we can say, "Sometimes a serpent is venomous."

The expression of quantity is also *often connected with that of quality*. In universal negatives the adjective "no" indicates both these elements of the predication. Thus we say, "No men are perfect"; which is probably a contraction for, "There are no men who are perfect." Occasionally, too, the particle "not" qualifies the quantity, and not the main matter, of the assertion. When we say, "All men are not wise," or, "Every man is not wise" ("all" and "every" being the emphatic words), we do not make universal predications, but mean, first, that some men are, or may be, wise, and secondly, that the men who are wise are not all. These modes of speech are based on the fact that the quantification of the subject is really a kind of predication which modifies that predication of which logicians speak; and they illustrate the general principle that the verbal proposition is often indirect in the expression of thought, and may need "interpretation."

9. Sir William Hamilton maintained that the predicate, as well as the subject, is quantified in every proposition. "The subject and the predicate," he says, "have each their quantity in thought. This quantity is not always expressed in language, for language tends to abbreviation; but it is always understood." On the basis of this doctrine and others connected with it,— as, for example, that a proposition is an equation, or an identification, of two notions, — Hamilton reconstructed all the formulas of reasoning.

The insuperable objection to Hamilton's doctrine is that it is not in accord with fact. It is not true that every predicate is quantified in thought. In ordinary categorical assertions the subject-notion is conceived as applicable either to all or to some of a logical class; in other words, is quantified; but *the predicate simply characterizes the subject.* When we say, "Birds are feathered; fishes live in the water; no metal is a vegetable; some men are trustworthy," we characterize the subject positively or negatively. It is no part of our thought that all feathered animals are birds; that only some things which live in the water are fish; that no vegetable is metal; and that no trustworthy beings are some men. The most that can be allowed is that the predicate is quantifiable. By reflecting on the nature of the assertion we can tell whether it gives information respecting the whole class which the predicate may name or respecting only a part of it. In the former case the predicate may be distributed; in the latter it is undistributed. But this quantification is no part of ordinary assertion; it is only an addition which may be made to it.

This is evident even in cases where the quantification of the predicate may easily take place; as, for example, in assertions which apply to all the members of a class and to them only. When we say, "All equilateral triangles are equiangular triangles," a little reflection on the relations of the things mentioned enables us to distribute the predicate, so as to say, conversely, "All equiangular triangles are equilateral." But if we did not reflect on the peculiarity of the case as affecting the predicate, the first proposition would merely signify that "Every equilateral triangle is equiangular." On the other hand, when we say, "Every equilateral triangle is half the rectangle formed by one of its sides and the perpendicular let fall on that side from the opposite angle," still more consideration is needed to perceive that this proposition *does not warrant the converse*, that "Every triangle which is equal to half the rectangle formed by base and perpendicular is equilateral."

10. Some teach that definitions always distribute their predicates, and that when we say, "Man is the rational animal," we mean, "All men are all the rational animals." But this is an extreme view. The object of definition is simply to characterize the subject definitely; which can be done without distribution of the predicate. Even the exact identification of the subject with the predicate belongs to the verbal form, rather than to the essential nature, of definition.

The only propositions which actually quantify the predicate are those which are intended to make assertions respecting the complete or the partial identity of classes. Such, especially, are those enumerations which identify a genus with some or all of its specific kinds. The predicate is distributed in saying, "Fishes, birds, reptiles, and mammals are the vertebrates"; it is undistributed in the statement, "The horse, the dog, the lion, and the tiger are some of the quadrupeds." As comparison of classes may take place in connection with any general statement, the predicate may always be quantified; but such comparisons occur only occasionally.

Here, however, "exclusive," or "exceptive," predications

should be mentioned, for they so qualify the subject as especially to suggest the distribution of the predicate. "Only men philosophize," or, "None but men philosophize," immediately suggests that all who philosophize are men. This mode of predication is really compound. The above example primarily asserts, first, that men philosophize, and secondly, that no other earthly beings do. But as our interest concerns those who philosophize and not the rest of the world, the inference arises unbidden that men — that is, some men — *are all the philosophers;* or that all philosophers are men.

Ordinarily the quantification of the predicate is not immediately suggested by the nature of the assertion, but is a special addition which prepares the original statement for the process called "conversion."

11. One other distinction between categorical propositions remains to be considered. It divides them into *the pure and the modal*. This classification was fundamental with Aristotle, and occupies a larger place in his system than the quantification of the predicate does in that of Sir William Hamilton; but it has been rejected by almost all logicians. It will be the first topic of our next discussion.

CHAPTER XII.

THE ILLATIVE PROPOSITION.

1. "Pure" categoricals might be styled "dogmatic." Verbally, they are assertions "*de inesse*"; and do not express logical sequence. 2. "Modal" categoricals are expressly either apodeictic or problematic. 3. Necessity, impossibility, possibility, and probability, as related to one another. 4. Logical necessity and possibility distinguished from causal. 5. Aristotle's modals wrongly rejected. 6. The distinction between "pure" and "modal" is verbal and superficial. 7. In addition to the six generic distinctions already made, propositions are (*a*) factual, or historical, and (*b*) illative, or inferential. 8. The illative assertion may be uncontracted in form; or contracted and categorical. 9. Except for usage, we might speak of "actualistic conditionals." 10. All inference may be expressed by two propositions. 11. Dogmatic propositions do not explain modal, but the modal the dogmatic.

1. A CATEGORICAL proposition is "pure" when, so far as verbal thought is concerned, it asserts the inherence or the non-inherence of the predicate simply as fact or truth. In other words, a pure categorical merely states that the predicate does, or does not, exist in its relation to the subject. "Arsenic is poisonous; some metals are not heavy," are pure categoricals, or assertions "*de inesse*": the first sets forth the existence of the quality "poisonous" in arsenic, the second the non-existence of "heavy" in some metals.

The term "pure" designates this class of propositions very inadequately. On this account, probably, Kant distinguishes them as "assertory." But all predications assert; the peculiarity of these is only that they assert simply, or without reference to any ground, or reason. A more distinctive designation is desirable: we may, therefore, sometimes style pure categoricals "dogmatic" statements, or dogmas; because, so far as verbal expression goes, they set forth matters of knowledge, or of belief, in a simple and unqualified way. The assertions, "Gold is valuable; savages are treacherous; all metals are minerals; some minerals are metals," may be styled dogmatic, because, in their simple positiveness, they resemble doctrinal statements, such as that God is merciful; that Jesus Christ died for sinners; and that the present life is one of probation for a future state of existence.

2. On the other hand, "modal" categoricals expressly present the existence or the non-existence of the predicate *as in some way logically connected with, or consequent upon, the existence of the subject*. They are of two principal kinds; the *apodeictic*, which predicates a thing either as necessary or as impossible; and the *problematic*, which predicates a thing either as possible, or as probable. The apodeictic has also been called the *demonstrative*, and the problematic, the *contingent*. "An unsupported weight must fall; animals cannot live without air," are apodeictic propositions; and two more such are contained in the sentence, "Straight lines parallel for any distance cannot meet however prolonged, but must continue parallel." "The straying horse may have taken any one of a number of roads," is a problematic categorical asserting possibility. "He has probably taken that leading to his former home," is a proposition of the same general class in which probability is asserted.

We must note that necessity, impossibility, possibility, and probability enter as elements into modal assertion *only because they indicate different modes in which the predicate may be logically connected with the subject*, and therefore, also, different forms and degrees of conviction regarding the predicate. To assert any one of these logical relations simply as a fact and for its own sake, and not as the ground for believing in something as necessary or possible or probable, would not be a modal predication. The statement that "proficiency in science is possible for any young American who avails himself of all his advantages" would be pure, not modal, were the design of the speaker simply to set forth the fact of the possibility. But should one say, "A young American who has used all his educational advantages may be — or is possibly — proficient in science," this would be a modal proposition; the possibility mentioned in it would be asserted, not for its own sake, but to indicate the basis and the character of a judgment.

3. Of the four modes of logical connection necessity and possibility are philosophically prior to the other two. Impossibility and necessity both relate to precisely the same state of things and differ only because of a difference in our modes of regarding things. Whenever anything is a fact, and no power can make it otherwise, it is necessary; and whenever anything is not a fact, and no power can make it a fact, it is impossible. But as, whenever anything is fact we can conceive also of that which is not fact corresponding to it, necessity and impossibility may always be asserted together. If it is necessary that something should be, it is impossible that it should not be; and if it is necessary that something should not be, it is impossible that it should be. Necessity is called positive or negative according as the fact to which it pertains is one of existence or of non-existence; and impossibility is characterized in the same way as belonging to that of which we conceive as the opposite of fact. Since impossibility thus originates from the same conditions with necessity, and is, as it were, the other side of the same thing, an understanding of necessity reveals the nature of impossibility, also.

Probability, likewise, is conditioned on possibility; yet not so simply and directly as impossibility is conditioned on necessity. When out of a number of possible alternatives one, and only one, must be true, and we have no reason to expect one more than another, we say that they are equally likely, or probable. But if a given proportion of those equal individual possibilities have a general character which may be realized in one and the same event, then that event is probable according to the ratio of the chances, or possibilities, for it, as compared with the total number of chances. Thus probability results from a general necessity combining with one or more individual possibilities of a given character. If a lottery contain ten blanks and two prizes, the drawing being settled as certain, the probability of any individual possibility is one-twelfth, and that of any given ticket gaining a prize is two-twelfths, or one-sixth. The connection of probability with possibility is indicated by the use of the same auxiliary verb, "may," to express either of these conceptions.

4. Logical necessity and possibility are of a very general character, and do not pertain only to effects as related to their causes. Therefore they must not be confounded with causal necessity and possibility. The latter are only specific modes of the former; for things may be possible or necessary without any reference to causation. It is possible that a pint of fluid should be contained in a quart measure; and it is necessary that the fluid which fills two pint measures should be equal to that which fills the quart measure. So, also, if an exterior angle be formed by prolonging one side of a plane triangle, it *may* be twice as large as one of the interior and opposite angles, and *must* be equal to both those angles taken together. In these cases necessity and possibility arise from arithmetical and geometrical, not from causational, relations.

5. The only modal propositions of which Aristotle treats are those which assert either necessity or impossibility or contingency; under which last head both possibility and probability are included. The Greek commentators on Aristotle, however, misapprehending his conception, enlarged the sphere of modality by making it embrace every proposition the predicate of which has an adverbial addition. According to them, "Alexander conquered Darius honorably," would be a modal. But this proposition differs only in expression from the more direct statement, "Alexander's conquest of Darius was honorable." Then subsequent logicians, perceiving that such adverbial propositions are in reality pure categoricals, concluded that a modal proposition is merely a pure proposition irregularly expressed. They were the readier for this doctrine, because Aristotle's discussions of those forms of argument in which modal thought is recognized, are marvellously complex and difficult. Thus it has come to pass that, at the present day, modal predication is barely mentioned by logicians, and is then immediately dismissed as of no scientific importance. "The whole doctrine of modality," says Professor Francis Bowen, "is now rightfully banished from pure logic."

But it is a mistake to suppose that modal does not differ

seriously from pure predication, or that the necessity and contingency which it asserts are simply parts of the ordinary predicate. The modality, like the "quantity," of assertions should be regarded as a sort of added predication of which the main body of the assertion is the subject; and which is intended to qualify and complete the assertion. As "all" and "some" do not modify our conception of the subject of a proposition, but only show whether the statement is universally or partially true of a class of things, so "must" and "may" do not modify our conception of the predicate, but only indicate whether the assertion is based on a necessary or on a contingent connection of the predicate with the subject. This is apparent when, instead of saying that, "Such a thing is necessarily or possibly or probably so and so," we say, "It is necessary or possible or probable that such a thing is so and so"; for, in this latter form of statement, the modal words evidently do not qualify the predicate, but only tell how the predicate is logically connected with the subject. Moreover, as all inference arises from perceiving the connection of things with each other, it seems unwise to deny the importance of modal assertions.

6. In order to understand the true significance of this class of predications, and of categorical propositions in general, it is necessary to note that the distinction between "pure" and "modal," like that between conditional and categorical propositions, is really of a verbal and superficial character, and must be supplemented by another distinction which relates to the essence of thought. For that logical connection which modal assertions express is often indicated by implication in pure categoricals, so that, were we to think of propositions only in their mental character, we must allow that many pure categoricals are of a modal nature. It is especially true that when a pure categorical embodies a general principle, it is modal in its inner meaning. Indeed, as a rule, pure universal statements are apodeictic in their force, and set forth necessary truths; while pure particular statements are problematic, and present principles of contingent belief. The assertions, "All men are mortal; no men are perfect; some merchants are successful;

some savages are not treacherous," when employed as premises in argument, signify that *man must die;* that *man cannot reach perfection;* that *a merchant may be successful;* and that *a savage may not be treacherous.*

7. The distinction pertaining to internal and mental propositions, to which we are thus brought, relates not only to those expressed in categorical form, but to all propositions whatever. It divides assertions into the *factual,* or *historical,* and the *illative,* or *inferential.* A factual proposition is one which asserts mentally what the pure or dogmatic proposition sets forth verbally, namely, some fact of existence or of non-existence, simply as such. "Cæsar conquered the Gauls; the Hindoos are Asiatics; Rome is in Italy; Locke was born in 1632," are statements of this character. On the other hand, any proposition which asserts something as a logical consequence, either expressly or by implication, may be distinguished as illative, or inferential. Factual assertions form the body of history or narration; illative constitute the most important part of philosophical knowledge and theory. Both are radical modes of rational thought; yet, of necessity, the enquiries of the logician have much more to do with illative than with factual statements.

8. To understand the scope of this fundamental distinction between modes of assertion we need not dwell longer on factual propositions; but a subdivision of illative predications seems necessary. For, in addition to the categorical method of indicating illation, which may be distinguished as *the secondary, or shortened, mode* (and which is of two species, the modal, and the pure, or the dogmatic), there is *the primary and uncontracted method,* one form of which has already been considered in the "conditional" proposition. For the conditional proposition is of a two-fold nature; it is not only suppositive, or hypothetical, but also illative, or inferential. This latter character is sometimes indicated by the word "then" introducing the consequent: instead of saying, "If the man be honest, he will prosper," we say, "If the man be honest, then he will prosper."

9. It is because of this inferential force that fully expressed hypothetical propositions have been styled "conditionals."

For a condition is not necessarily a thing supposed. Ordinarily it is that which is requisite to the existence of anything. In the present connection it signifies the reason, or logical antecedent, of some consequent. For this always either is, or contains, the only condition requisite for the existence of the consequent. Other conditions than the one contained in the antecedent may be requisite, and often are, but in that case these are assumed as already existing, so that the consequent depends on that one condition only, and must follow if that condition be a reality. In order that the man may prosper, other things than honesty may be needful; these, however, are known or assumed when we say, "If the man be honest, he will prosper." But evidently a logical antecedent with its necessitant condition, no less than the ordinary and merely necessary condition, may be either real or supposed. Therefore the restriction of the term "conditional" to hypothetical statements is somewhat arbitrary.

Therefore, also, we say that there is another primary expression of illation, which, were it not for fixed usage, might be called a conditional proposition. We refer to *actualistic assertion when it is made on the strength of some given reason*. For example, "Since the man is honest, he will prosper; because the night has been clear, there must be dew on the grass; the triangle is equilateral, for it is equiangular," might be styled actualistic conditionals. But, because their conditions, or antecedents, are not suppositions, but realities, we must call them uncontracted actualistic illatives.

The question may be raised whether an uncontracted statement of illation can properly be called a proposition, inasmuch as it contains two propositions, one of which is inferred from the other. This question relates both to "conditional" propositions, and to those actualistic assertions just considered. We reply that these uncontracted statements may be styled either inferences or propositions according to the manner in which we view them. If our interest be chiefly directed to the thing inferred, then *the consequent assertion together with the antecedent as a kind of prefix*, is called a proposition; but if our scrutiny regard antecedent and consequent alike, then we speak of an inference. At the same time it may be allowed that any uncontracted statement of an inference contains much more than a mere proposition, and receives this name only through a secondary use of language. On the other hand, the illative categorical, whether modal or dogmatic, is rightly called a proposition, because, so far as verbal form is concerned, it is a single existential statement; and because our language relates primarily to the verbal form. Mentally, every such proposition is an inference which has the subject of the categorical for its antecedent and the predicate for its consequent.

10. In speaking of a certain style of proposition as the uncontracted expression of an inference, we do not mean that inferences are always expressed by it in the fullest possible way. Philosophical completeness often calls for developed forms of statement in which things are expressed which would otherwise be understood. We mean only that ordinarily and primarily an inference is stated by two propositions in the sequence of reason and consequent.

Moreover, this binary combination of propositions *can* be so used as to state any inference completely; and is the only form in which every inference may be completely stated. When the antecedent is a single truth or fact, the inference must be set forth by two propositions; and when the antecedent is a combination of assertions, this combination must be regarded as one complex statement, to be followed by one other statement as its consequent. "If all men are mortals, then some mortals are men," is an inference which admits of only one premise; while in the following inferences a plurality of premises is expressed by one statement: "A is equal to B which is equal to C; therefore A is equal to C. If A is older than B who is older than C, then A is older than C. Since A is a part of B which is a part of C, A is a part of C. Because Hindoos belong to the class Men who are mortal (or have the nature of man which necessitates mortality), Hindoos are mortal." All inference, therefore, may be expressed by one antecedent proposition and one consequent. This is as it should be; for it is the essential nature of inference to assert a consequent because of a reason.

11. Since the illative categorical proposition is a secondary and shortened form of statement, it evidently should be explained by a reference to the primary and uncontracted inferential proposition. This, however, reverses the ordinary teaching of logicians. For, after entirely discarding modal predications, they explain "conditional" propositions as being of the same nature with pure categoricals; and then base their theories of reasoning on the recognition of these last alone. The unsatisfactory character of this course is especially apparent when we see how artifice is often needful to give categorical form to a conditional statement. Sometimes this is effected without difficulty; the assertion, "If iron is impure, it is brittle," is easily replaced by, "Impure iron is brittle." But artifice is often necessary; as, for example, when recourse is had to the words "case of." Thus the conditional, "If Aristotle is right, slavery is a proper institution," is transmuted into the categorical, "The case of Aristotle being right is the case of slavery being a proper institution." This statement is both less natural and less explicit than the original proposition. In particular the "case" might be a real instead of a supposed case, and instead of signifying, "If Aristotle is right," and so on, might signify, "Since Aristotle is right, slavery is a proper institution."

It is frequently taught also that the conditional judgment, when transmuted, always gives rise to an universal categorical. This is a mistake resulting from the fact that most conditionals assert a necessary consequence. Cases of contingent sequence cannot be expressed by a pure universal proposition. "If iron be brittle, it may be impure," has for its modal equivalent, "Brittle iron may be impure," and for its dogmatic equivalent, "Some brittle iron — or brittle iron sometimes — is impure."

The consideration of illative assertion concludes the second part of logic, which relates to propositions, or existential statements, as such; and it has had the effect of introducing us into the third part of logic, which concerns inference. This result has arisen from the duplex character, verbal and mental, which belongs to propositions as expressions of thought. Primarily propositions express assertion, simply, and not inference.

CHAPTER XIII.

INFERENTIAL SEQUENCE.

1. Inference defined. The law of Reason and Consequent. 2. Inferences are (*a*) single-grounded, double-grounded, or many-grounded; (*b*) immediate or mediate; (*c*) apodeictic or problematic. 3. The law of Conditions. 4. Necessary, or logical, relations. 5. A condition does not necessitate, but is necessitative. The exact necessitant, which reciprocates with its consequent. 6. Why the ordinary necessitant does not reciprocate. 7. The inference of the necessary to be, and of the necessary not to be, or the impossible. 8. Possibility: its nature and modes; the law of its inference. 9. The possible not to be; how inferred. 10. Contingency. 11. Probability.

1. INFERENCE, or illation, is the process whereby we assert one thing to be fact or truth, because of its connection with some other thing, or things, which we assume to be true. The radical principle to which all inference conforms has been called the law of Antecedent and Consequent, or of Reason and Consequent, or of Sufficient Reason, or of Adequate Reason. This law simply generalizes the truth that two things, or facts, by reason of their respective natures, are often so connected with each other that the reality of the one may be the ground of our believing in the reality of the other. Inference, therefore, takes place, not from the principle of Antecedent and Consequent, but only according to it. In every case we think first of a reason and then of a consequent; but the consequent is accepted because of its connection with the reason, and not because it is connected with the law of Reason and Consequent. So far from individual inferences depending on a knowledge of this law, the knowledge of the law is derived from analyzing them.

2. An inference, if based on a single fact or statement, may be called single-grounded; if based on two statements it may

be styled double-grounded; and it may be distinguished as many-grounded, if based on three or more statements. But even though there may be a plurality of premises, there is only one antecedent; and for the constitution of this the premises must be combined.

Inferences are either immediate or mediate. In the former a consequent is inferred from an antecedent without any intervening link of illation. We assert that because A exists, C exists: we say, "Because this line is straight, it is the shortest possible between the terminal points." On the other hand, mediate inference arises when the consequent of a first inference is the antecedent of a second; and it conforms to the law that *the antecedent of a second antecedent is the antecedent also of the second consequent.* Let B follow from A, and C from B; then C follows from A. We say, "If he is human, he is rational, and if he is rational, he is responsible; therefore if he is human, he is responsible." Such inference is mediate because a second inference intervenes between the first and the third; and yet more because, in the concluding inference, the antecedent of the first inference becomes connected with the consequent of the second through that common part which is consequent of the first and antecedent of the second.

Mediate inference is often called reasoning, or ratiocination; while the word "inference," when used alone, generally signifies immediate inference. In the remainder of the present chapter, using the term in this limited signification, we shall discuss immediate inference only. The consideration of this topic properly comes before that of reasoning.

Viewed with reference to the mode of sequence between antecedent and consequent, inferences may be divided, as illative propositions have been, into *the apodeictic and the problematic.* In the former of these consequents are inferred as necessary, or as impossible; in the latter as possible, or as probable. This distinction is sometimes expressed by saying that inferences are either necessary or contingent. In one sense, however, all correctly formed inferences are necessary; for every just conclusion is necessary as a matter of belief, or conviction; even though it may not be necessarily, but only possibly or probably, true. The principles of necessary conviction may be either problematic or apodeictic, while those of necessary truth are apodeictic only. It is, however, most important to note that *every mode of inference arises from a recognition by rational beings of the necessary relations of things.*

3. The significance of this statement will become apparent if we can perceive how the sequence of antecedent and consequent in its various forms is connected with an universal law of existence, namely, that *all beings and natures exist under conditions, every nature having conditions of its own, and like natures having like conditions.* This law, which, like that of reason and consequent, is the generalization of a necessity perceived in individual cases, may be styled the law of Conditions. It can be easily apprehended, provided only we carefully determine that conception which philosophy here attaches to the word "condition."

This term, derived from "condere," to put together, or construct, first signified whatever may be connected with the formation, or constitution, of a thing; any prerequisite, or constituent, or concomitant; then it was used, more widely, for whatever may attach itself in any way to the existence of a thing. Thus industry is a condition of success; sanity and insanity are conditions of mental life; the fulfilment of the terms of a contract is the condition of a claim to its benefits; one's financial condition is the state in which he finds himself as related to pecuniary resources.

In the loose general sense now described, conditions may be either contingent, and occasional, or necessary, and invariable. For, while industry is the indispensable condition of success, sanity is not inseparably connected with mental life; and one who is now in a bad financial condition may be prosperous hereafter. Even the stipulation of a contract may be so changed or supplanted that the performance may be no longer required before the bestowment of the reward. Philosophy and logic, however, speak of those conditions only which are absolutely necessary to the existence of a thing; therefore we now define a condition to be *a second thing so related to a first that the first cannot exist without the second;* from which it follows, also, that if the first thing exist, the second must exist along with the first. Space is a condition of motion; because without space there could be no motion, and if there is motion there must be space.

4. The relations connecting conditions with the things conditioned are of great variety. They pertain to objects as in space or in time, or as having number or quantity or size or shape, as being numerically the same or different, as being similar or dissimilar in nature, as being wholes and parts, or as being endowed with active and passive qualities and subject to the law of cause and effect. The metaphysician studies such relations specifically; logicians are concerned only with their general character as being necessarily connected with the conditioned object, and as necessitating the condition.

In saying that the condition is necessitated by reason of its relation to the thing conditioned we refer to logical, and not to causational, sequence. In logic an effect may necessitate a cause as truly as a cause an effect. God, as eternal and self-existent, is free from causal conditions and causal necessity, but, as creator of the universe, He is logically necessitated. He is the only adequate, and, therefore, the necessary, cause of the universe. Then, as we have seen, necessity may arise from other relations than those of cause and effect. When we say, "What is part of a part is part of the whole," we do not present one fact as caused by another, but only as accompanying that other in a way that no power could prevent or change. That general necessity of which logic speaks has an all-comprehensive sphere; it is well described by Aristotle when he says that *when a thing exists, and no power can make it not exist, it is necessary.* For he thus teaches that necessity does not originate from power, but is the quality, or relation, which belongs to certain modes of existence as being beyond the operation of power.

While a condition is necessitated by the thing conditioned, and may be inferred from it as a consequent from an antecedent, it is noticeable that we seldom regard the same thing as a consequent and as a condition   For we speak of conditions when our enquiry concerns the thing conditioned, but of consequents when the question concerns the consequent itself. Consequents differ from ordinary conditions because they belong to that class of objects which secures our primary interest and attention; and to which things which we conceive of as conditioned also belong.

5. If things could be inferred from their conditions, as these may be inferred from the things conditioned, we would never be at a loss to find an antecedent for a consequent. Such is not the case. Nevertheless, while a condition ordinarily does not necessitate what it conditions, it has what may be called a contributory necessitative force, and by reason of this it may help to necessitate what it conditions. For example, a straight line is a condition of a plane triangle, and must exist if the triangle exist; but it does not necessitate the triangle. If, however, there be three straight lines of indefinite length in the same plane which do not cross one another at the same point, and no two of which are parallel to each other, there must be a triangle. Thus three or four conditions may so combine as to form an antecedent necessitating the thing conditioned. Such a combination of conditions is itself a compound condition; it is both necessary as a condition and necessitating as an antecedent; therefore it may be styled a necessitant condition.

Whenever an antecedent is thus a necessitant condition of its consequent, the inference admits of simple conversion. In other words, the consequent may be used as an antecedent, and the antecedent as consequent, in a new inference. For this reason, in reference to a four-sided rectilineal figure, we can say, not only, "If the opposite sides are parallel, the opposite angles are equal," but also, "If the opposite angles are equal, the opposite sides are parallel." And, if either of these things be not so, we can say that the other is not so too.

6. Ordinarily, however, an antecedent is not simply a necessitant condition, but is *something which contains such a condition*. Therefore most inferences do not admit of simple and thorough conversion. The logical rule is that we may assert the reason and then assert the consequent, and that we may deny the consequent and then deny the reason; but that we

cannot either assert the consequent and then assert the reason, or deny the reason and then deny the consequent. The ground for this rule is that although the consequent cannot exist without such a necessitant condition as is contained in the antecedent, such a condition may be contained in some other antecedent; and so the consequent may exist while that antecedent does not exist, and that antecedent may be non-existent while yet the consequent is a fact.

The nature of a necessitant condition may be illustrated by that of the exact philosophical cause of an effect. Sometimes we say that the same effect may be produced by a variety of causes; and this is true. One may become warm by exercise, or with a fire, or by putting on heavy clothing. Yet there is a common heart, or core, in each of these methods, on which its efficiency depends: in each case the heat arises from the collection of a certain amount of chemical, or molecular, action upon or within one's body. So, also, the general cause of disease is the partial failure of some bodily function; and this failure may take place in a variety of ways. The general cause of motion is the uncounteracted application of force; but force may be either attractive or repulsive, and may be exerted by either animate or inanimate agency. The philosophical cause is logically convertible with its effect; but this is not true of those various causes in each of which the philosophical cause is wrapped up. In like manner, every logical antecedent contains within itself a necessitant condition of its consequent, and derives its life, or illative force, from that condition. Such exact and convertible necessitants occasionally present themselves in our reasonings, especially in the demonstration of mathematical theorems. Ordinarily, however, careful discrimination is required to dissect them out of their envelopments. The discovery of them is the work only of philosophical thought. Yet even the simplest antecedent, if it be not itself a necessitant condition, can be shown to contain one. The consequence, "If there be motion, there is space," does not yield the converse, "If there be space, there is motion"; nor is it easy to say at once what element or property of motion is a necessitant condition of space. But analysis answers this query. For motion involves increase or diminution of distance; now where there is distance there is space, and where there is space there is distance.

7. We can now state explicitly how the different modes of logical sequence are related to, and based upon, the law of Conditions. The existence of an entity is inferred as necessary when an antecedent either is or contains a necessitant condition. This is the law of positive apodeictic inference, or of the perception of the necessary to be.

The inference of negative necessity — or of the necessary not to be — is yet more simply related to the law of Conditions. For a thing is necessarily non-existent so long as any of its conditions are non-existent. When, therefore, an antecedent either asserts or involves the non-existence of some condition of an entity, the non-existence of the entity is a necessary consequent.

Along with the necessary non-existence of an entity, and from the same antecedent, we can infer the impossibility of its existence; and along with the necessary existence of a thing, and from the same antecedent, we can infer the impossibility of its non-existence. For negative necessity and positive impossibility, as also positive necessity and negative impossibility, differ only as being different sides, or aspects, of the same consequence.

But the impossibility ordinarily mentioned is the impossibility to be. For the human mind finds that it is easier and pleasanter to form a positive conception and then to reject it as incompatible with the antecedent, than it is immediately to conceive and to assert a thing to be necessarily non-existent. In like manner the necessity commonly mentioned is the necessity to be.

Passing from apodeictic to problematic sequence we ask, "How are the inferences of the possible to be and of the possible not to be, related to the law of Conditions?"

8. The possibility, like the necessity, of a thing is a relation between the existence of it and that of other things. When a thing is necessary, its existence is absolutely coherent with that of other things: when it is possible, its existence is

compatible with that of other things. This compatibility may belong either to an actual or to a supposed object, and may, therefore, itself be either actual or supposed. The suppositional mode of possibility is that thought of when we infer a thing as possible; for such inference is useful only when we do not know that the thing *is*, and yet can say that it *may be*.

Possibility, whether belonging to an actual or to a supposed object, admits of degrees, the lowest of these being that existential consistency which one thing may have with others, with which it has no special natural connection. Thus it is possible that there should be a man, or a house, or a tree, upon a prairie. This degree of possibility is so far removed from proof, and even from suggestion, that it scarcely has a place in logic. Ordinarily, the possibility of an entity — its existential compatibility with given circumstances — is more than mere consistency. It implies that *the circumstances contain one or more of the proper, or special, conditions of the entity, and that, in this way, they present a suitability for its existence*. When we say that, under such and such circumstances, a thing is, or would be, possible, we commonly mean, not simply that the circumstances would admit of the existence of the thing, but that they are specifically compatible with its existence, because they contain one or more of its necessary conditions.

When an entity really or necessarily exists, all its conditions exist; it is compatible with all its circumstances; it is possible in every respect, or in the highest degree. But this thorough-going compatibility, being recognized only when a thing is already perceived as real or as necessary, is seldom used as a ground of inference. And so it happens that the ordinary possibility of logical sequence is neither the weakest nor the strongest possibility, but is intermediate between the two. It is that conceived compatibility which arises upon our perceiving one or more of the conditions of an entity, while other conditions are not yet known either to exist or not to exist.

This possibility may co-exist, or is consistent, with either necessity or impossibility. For, on the one hand, the discovery of conditions not yet known to exist may enable us to form a logical necessitant; and, on the other hand, if investigation shows some condition to be excluded from the given circumstances, the thing is impossible; that is, it is necessary not to be.

Hence it is evident that *an entity is inferred as possible to be when an antecedent*, without justifying an apodeictic conclusion, either positive or negative, *is, or contains, one or more of the conditions of the entity*. In other words, the existence of a thing is inferred as possible when some of its conditions are known to exist, while the rest are not known either to exist or not to exist.

9. The possible not to be is inferred from precisely the same antecedent as the possible to be, but the parts of the antecedent are used differently. This might be expressed by saying that *the non-existence of a thing is inferred as possible when some of its conditions are not known either to exist or not to exist, while some are known to exist*. The reason for the first part of this statement is that when circumstances are not known to contain some conditions of an entity, the non-existence of the entity is compatible with the circumstances so far as they are known; the reason for the second part is that we have no inducement to enquire concerning the possibility of non-existence except in cases which suggest the possibility of existence. The whole doctrine concerning both the inferences of possibility may be summed up in the statement that the possible, either to be or not to be, is inferred when some of the conditions of a thing are known to exist and some are not known either to exist or not to exist. But we must add that the inference of negative possibility is comparatively infrequent, and that, generally, the possible means the possible to be.

10. We have seen that the possibility on which inference is based is a compatibility, and not a mere consistency. When this compatibility is so specific in its conditions as to be of a decided and noticeable character, it gives rise to a judgment of *strong possibility, or of what has been called contingency*. For we do not say that it is contingent to a man, or even to a man of talent, to write poetry, but to a poet: because, in this last case only, there is a special adaptedness for that work. This contingency may be said to approach necessity, for it always

suggests that an apodeictic antecedent may be found. Were the question whether a certain man made a well-fitting coat or not, and it were ascertained that he was in the tailoring business, the contingency thus arising would stimulate the search for a necessitating reason; though in itself it would only justify a judgment of strong possibility.

This brings us to the last mode of logical sequence, namely, the inference of a thing as probable; for this inference holds a place intermediate between those of contingency and of necessity.

11. Every inference that an entity is possibly, or contingently, existent, may be accompanied by another, based on the same data, that it is possibly, perhaps contingently, non-existent. Neither of these inferences, however, results in a definite confidence that the existence, or that the non-existence, is a fact; we only say that each of these consequents is contingent or possible, and that, therefore, there is nothing absurd or unnatural in the supposition of it. Contingency, indeed, — the strengthened form of possibility — is accompanied with some expectancy; but this is of an entirely weak and indeterminate character.

But, as it is certain that a thing must be either existent or non-existent, it is plain that the confidence of certainty may, in any case, be definitely divided between the positive and the negative possibilities, *provided only we can determine what share belongs to each*. Now this apportionment of confidence takes place whenever an antecedent of possibility, or of contingency, becomes so modified that it must be followed by some one of a number of events which are equally possible, and when the consequent enquired about is either one of these events, or is of such a nature as to agree with more than one. A drawing from a collection of variously colored balls in indefinite and unknown numbers, would be an antecedent of possibility with reference to the appearance of any individual ball first, or even of a red or of a white ball first. But, if we were informed that there were just thirty balls, twenty white and ten red, then — *since the antecedent as modified by this knowledge gives the same amount of bounded, or limited, contingency, or expectant possibility, for each ball* — we say that the probability for the appearance of any individual ball is one-thirtieth, while that for a white ball is twenty-thirtieths, and that for a red ball, ten-thirtieths.

The individual events, or consequents, conceived of and inferred as equally possible, and as the only possibilities in the case, and the number of which is the denominator of the fraction of probability, are called chances.

Because the inference of probability has much in common with that of contingency, and may even be regarded as a modified inference of contingency, and because the same verbal forms are used to express both, they have often been classed together and called contingent inference. But they should not be confounded with one another. Both modes of inference will be discussed more fully hereafter.

CHAPTER XIV.

ORTHOLOGIC INFERENCE.

1. Inferences are also orthologic or homologic. 2. We infer, primarily and ordinarily, not from, but according to, the ultimate laws of inference. 3. One individual fact may be orthologically inferred from another. 4. The principles of orthologic inference are (*a*) logical, or universal; (*b*) semi-logical, or specific. 5. The logical laws are (*a*) those of identity, contradiction, and excluded middle; which relate to the existence and non-existence of things; and (*b*) axiomatic principles concerning the common accidents of entity. 6. The semi-logical are (*a*) metaphysical axioms, (*b*) mathematical. 7. In orthologic inference (*a*) scrutinize the antecedent, (*b*) formulate the law. 8. The principle of identity is the unchangeableness of fact or truth. 9. It justifies (*a*) definitional substitution, (*b*) the synthesis of assertions, (*c*) the conversion of propositions, (*d*) analytic and subordinative judgments. 10. The law of contradiction supports (*a*) the rejection of absurdity, (*b*) the avoidance of inconsistency, (*c*) the *reductio ad impossibile*, (*d*) "contrapositive" inference. 11. The law of excluded middle is logically prior to the other two; and is used mostly in combination with the law of contradiction. This combination was Aristotle's "first of first principles."

1. Some inferences attach their consequents to their antecedents without referring to any previously known case of existential connection. Others, referring to some previously perceived case of necessity or contingency, base their validity on the similarity of the antecedent now presented to that formerly perceived. Let us term those inferences whose validity depends on this reference, *homological;* and those whose force is independent of any previous perception of connection, or consequence, *orthological*. Both these modes of illation take place in accordance with law; but they differ in that orthological inference follows a considerable variety of laws, while homological inference is based on that one law which unites like consequents, whether of necessity or of contingency, with like antecedents.

2. The laws of inference when definitely formulated are termed "principles" of conviction; and such of them as cannot be resolved into simpler laws are called "ultimate" principles. It was formerly taught that the mind has the power of immediately perceiving these ultimate principles, and that all inference and reasoning depend on the application of them as rules to specific cases. Beyond question we sometimes reason in this way; and therefore, because of the order of our thought in such reasoning, the ultimate have also been styled the "first" principles of conviction. Nevertheless the doctrine, obscurely taught by Aristotle and more thoroughly advocated by Locke, that all our knowledge originates in the perception of particular facts and cases, and that general notions and principles are derived from individual perceptions by a process of analysis and abstraction, is indisputably true. First principles are "first" only as principles, or rules, not as perceptions; and they are styled "self-evident" only because they are immediately and easily obtained from individual perceptions, and require no proof except that they be illustrated and tested by a reference to such perceptions. For while an axiom shows what elements in a case render a certain consequent necessary, it adds nothing to the certainty of the inference, and it may be unthought of, and even unknown, while yet one is reasoning in accordance with it.

"I ask," says Locke, "is it not possible for a young lad to know that his whole body is bigger than his little finger, but by virtue of this maxim, that the whole is greater than a part, nor to be assured of it till he has learned that maxim? Or cannot a country wench know that, having received a shilling from one that owes her three, and a shilling also from another that owes her three, the remaining debts in each of their hands are equal? Cannot she know this, I say, without she fetch the certainty of it from this maxim, that, if you take equals from equals, the remainders will be equals, a maxim which possibly she never heard or thought of? I desire any one to consider . . . which is known first and clearest by most people, the particular instance or the general rule; and which it is that gives birth and life to the other." (Essay, Bk. iv. 12.)

3. That we constantly reason without the use of first principles as rules should be especially borne in mind in connection with orthologic inference. For this, primarily, is *the inference of one individual fact from another with which it has some special necessary connection.* The cases of inference mentioned by Locke are orthologic, and evidently, though conforming to general principles, they do not depend upon them; nor do they depend upon any previously perceived case of similar sequence.

Probably, as a matter of fact, the necessary relations of entity, together with the relata which they connect, are first perceived presentationally, and only afterwards are employed in inference. But our inferences concerning things as thus related, contain no reference to any such previous perceptions.

Moreover, when conceiving of these relations, we recognize them, not merely as parts of an established or ordained constitution, but as absolutely necessary, and as belonging to that nature which things must have if they exist at all. Orthological inference, therefore, as being specially related to the unchangeable constitution of things, is, in a pre-eminent sense, ontological.

4. The classification of inferences is naturally the same with that of the principles on which they proceed, every principle being the formative law of the inferences corresponding to it; moreover any mode of inference is best explained by stating clearly the principle according to which it takes place. Orthologic principles may be divided into two grand classes, which, for want of better terms, may be distinguished as the *logical,* or *universal,* and the *semi-logical,* or *specific.* The former pertain to all entites whatever; the latter to different radical forms of entity, as such. We call the one class of principles logical, because the modes of conviction to which they give life are of unrestricted applicability, and are discussed in the general science of reasoning; the other class of principles are semi-logical, because the logician, though distinctly recognizing their illative force, is not concerned about their specific nature and workings.

5. The universal laws may be subdivided into two classes.

The first of these concerns *the existence and non-existence of things;* and is composed of the three important laws of *identity, contradiction* and *excluded middle.* We shall endeavor, presently, to explain these laws. The second class relates to all entities *as having certain common "accidents," or properties, namely, individuality, quantity, and character, or nature.* From these properties and the relations founded upon them, arise number, and numerical identity and difference; also specific character, and identity and difference in kind; and also the conception of whole and parts; whether of the metaphysical whole, or substance, and its attributes, or of the logical whole, or class of similars, and its members.

The axiomatic principles pertaining to these universal aspects of entity are such as the following: everything that exists must be an individual, and is numerically different from other things and numerically identical with itself, so that we can say "it is this, and not that": — every entity has a nature of its own, in which, however, it more or less agrees with, or resembles, other entities; so that it may be enrolled now in this logical class, now in that one: — every entity may be regarded as a metaphysical whole, or substantum, with attributal parts: — whatever is included in, or connected with, an attribute, is included in, or connected with, the whole thing, or substance; but what is inconsistent with any attribute is inconsistent with the substance: — what is true of a class of entities distributively must be true of every subordinate class or individual: — when two things are each identical with a third, they are identical with each other; but when one is identical with a third and the other is different from it, they are different from each other: — in like manner, if each of two things agrees with a third in having some character or nature, they agree with each other; but if one agrees and the other disagrees, they disagree with each other. To these axioms, or laws of necessity, postulates, or laws of possibility, might be added; for example, what consists with an attribute may consist with the substance (or "substantial form"), and what is true of a specific class may be true of the generic.

These and other universal principles, which relate to the

common nature, rather than to the existence and non-existence, of entities, are interwoven with the very structure of human thought, and are the basis of important logical operations. Yet they are so simple, and so unobtrusive in their operation, that they are not often discussed at any length, but only referred to as self-evident.

6. The second grand class of orthologic principles, the semi-logical, are those which support reasonings respecting specific modes of Being. They, also, may be subdivided into two classes; the *metaphysical* and the *mathematical*. The former control our judgments respecting the most generic kinds of entity; the latter, our specific reasonings regarding the quantitative and spatial relations of things.

Metaphysical axioms or laws, are such as these: Space and Time exist: — all other entities exist in space and in time: — Space and Time, though conditions of production and destruction, cannot themselves be produced or destroyed: — every body occupies space: — two bodies cannot occupy the same space at the same time: — no body can be in two places at once: — no body can successively occupy two separate locations without passing through the intermediate space: — all powers reside in substances, and are exercised by substances only: — every beginning or change is the result of the exercise of some power: — power acts only on or in substance: — power never acts without conditions, and the exercise of a power, together with its necessary conditions, constitutes a cause: — a cause and its effect (that is, the change consequent upon the cause) are inseparably united, so that neither can be present or absent without the presence or absence of the other: — every change corresponds in its nature to the cause producing it: — where there is no cause for a change, things remain as they are: — the cause of a cause is the cause of the effect: — a part of a part is part of the whole: — what resembles a likeness resembles the original: — what excludes, or contains, a container, excludes, or contains, its contents.

Mathematical axioms are such as these: Space admits geometrical figures and relations: — quantity admits of measurement and its relations: — a whole is equal to the sum of its parts: — a whole is greater than any of its parts: — a straight line is the shortest possible between two points: — through a given point one, and only one, straight line can be drawn parallel to a given straight line: — a straight line may meet another straight line so as to make two, and only two, equal adjacent angles; and all the angles so made (that is, all right angles) are equal to one another: — angles, and other magnitudes, which can be made to coincide with one another are equal: — solids of similar shape are equal if their boundaries are equal: — if a first thing be equal to a second which is equal to a third, the first is equal to the third: — if a first thing be greater than a second which is equal to, or greater than, a third, the first is greater than the third: — magnitudes of the same kind must be related to each other as equals, or as the greater and the less: — if A equal B, and C equal D, and if A be added to C, and B be added to D, the sum of A and C will equal the sum of B and D: — two straight lines parallel throughout any part of their course, will continue parallel however they may be prolonged.

7. Such are the various classes of orthologic principles. To reason correctly in accordance with these principles requires care and thoughtfulness, but does not call for much artificial guidance. The act of inference, in itself, is very simple. Antecedent and consequent being considered in their relations, the latter is immediately asserted. The principal rule to be observed is that we should exercise careful scrutiny so as to determine what the antecedent presented may be, and whether it be adequate or not: and in this work it will help us if we state the inference in general terms and formulate the law on which its validity depends. Ordinarily this law can be ascertained without difficulty.

8. Let us now recur to that triad of principles which relate to things simply as existing and as non-existent; for, while the importance of these laws is beyond dispute, the nature and use of them have not always been clearly apprehended. Let us note, first, that *the principle of identity pertains to facts or*

statements which are identical with one another, and not to facts or statements of identity. That is, it no more pertains to these latter than to any other facts or statements. This principle asserts the unchangeableness of fact and truth. It has been expressed objectively by saying, "Whatever is, is; and whatever is not, is not," and subjectively by saying, "Whatever is true, is true; and whatever is not true, is not true." These maxims are needless and useless as grounds of deductive inference; but they are fundamental laws of thought. They compel us either to abide by any statement already made or to confess that we have not spoken the truth; and they require us to accept a true statement a second time, or any number of times, even though it should be accompanied by non-essential additions, or modifications.

9. The *right to substitute the definition* of a name or notion for the name or notion depends on the law of Identity. Common salt being chloride of sodium, it is an orthologic inference to say, "Good health involves the use of chloride of sodium, because good health involves the use of common salt."

Again, the principle of Identity is employed *when we combine two statements respecting the same subject; or unite equivalent modifications to both extremes of a proposition; or join any two statements of a congruous nature, so as to make one compound assertion.*

    Gold is a metal;
    Gold is valuable; therefore,
    Gold is a valuable metal. —

is an inference made by combining two statements respecting the same subject. A precisely similar combination occurs in that transformation of thought which we call the substantialization of the predicate. Thus,

    Gold is a thing;
    Gold is valuable; therefore,
    Gold is a valuable thing, or a valuable.

The following inferences result from adding equivalent modifications to both terms of a proposition:

    A negro is a fellow-creature; therefore,
    A negro in suffering is a fellow-creature in suffering.
    Oxygen is an element; therefore,
    To obtain oxygen is to obtain an element.

The union of congruous statements yields such inferences as the following:

> Industry deserves reward; and
> A negro is a fellow-creature; therefore,
> An industrious negro is a fellow-creature deserving reward.

Any synthetic statement may be justified when thus compounded of assertions which are individually correct.

In the next place, *that inference from substantal predications which logicians call conversion*, is based on the law of Identity. Every substantal predication either asserts or denies the identity of its subject with its predicate; its converse makes the same assertion, though with a variation in the order and emphasis of thought. The predicate of any proposition having been, if necessary, substantialized and quantified — for example, "all men are mortal, being made "all men are some mortals" we immediately say, "some mortals are men — or all men."

The conversion of substantal predications is that commonly mentioned, and is of special logical significance; but any relational assertion may be converted in a similar manner. The inferences,

> William is the husband of Anna; therefore,
> Anna is the wife of William:
> A is equal to B; therefore,
> B is equal to A —

proceed on the principle of Identity.

Finally, *this principle may be used to justify the analytic, and the subordinative, judgments*. In the former of these we predicate an attribute of a subject of which we already know the definition; this predication may be considered a partial repetition of the definition. In the latter we assert of *some* what we know to be true of *all*, and this may be considered a repetition in part of the universal statement; because, in thinking of the all, we may have thought of the some also. If, however, it be objected that we do not always at first think of the some as being in the all, and of the attribute as being in the essence, this may be allowed. In that case the inference in question would follow, not the law of Identity, but principles relating to the metaphysical and logical wholes. For any part of an essence must be an attribute of the substance; and what belongs to all must belong to some or any. This latter axiom is the dictum of Aristotle.

10. The principle of Identity which compels us to maintain what we have learnt to be true, and to deny what we have ascertained to be false, operates in the mind more constantly than any other law of inference. But this operation almost evades our consciousness; it is so easy and spontaneous. The law of Contradiction, on the contrary, is so frequently used for the rejection of error and the confirmation of truth that it was held by Aristotle to be the first of all first principles. This law asserts that the presence of existence and the absence of non-existence — as also the absence of existence and the presence of non-existence — involve each other. Objectively, it says that the same thing cannot be and not be at the same time, but must either be or not be. Subjectively, it says that when a proposition (positive or negative) is true, the contradictory of it is false, and that when a proposition (positive or negative) is false, the contradictory of it is true. The first part of this law governs immediate contradictory denial, the second immediate contradictory affirmation. For the principle relates only to that contradiction which may take place when one proposition sets forth the existence, and another the non-existence, of the very same thing.

The chief use of the principle of Contradiction is indicated by its name. It enables us to assert that the opposite of what we have found to be true is false, and that the opposite of what we have found to be false is true. It so links together what is fact and what is not fact, what is true and what is not true in any respect regarding any subject, that, when either of these is known the other may be known also. Let "due" mean "not paid"; then the debt, being paid, is not due; being not paid, it is due; being due, it is not paid; or being not due, it is paid. So also we might contrast "present" and "absent."

Because this law prevents us from believing in two opposite things at once Sir William Hamilton styles it the principle of *non-contradiction;* but the older name is to be preferred as giving the immediate effect of the law; which is the rejection of error as the opposite of truth, and the assertion of truth as the opposite of error.

A specific use of the principle of Contradiction occurs in that method of argument known as the "reductio ad absurdum," or "ad impossibile." If the immediate contradictory of an assertion be false, the assertion must be true. Let us assume that contradictory as an antecedent and show that it leads to a false conclusion. This being done we say that the assumed contradictory must be false, and therefore, also, the original assertion true. For any antecedent which necessitates a false consequent must itself be false. That a straight line cannot meet the circumference of a circle in more than two points is proved as follows. "For *if it could* meet it in three or more points, all those points would be equally distant from the centre, and hence there would be three or more equal straight lines drawn from the same point to the same straight line. But this is impossible. Therefore the antecedent, contradictory of the original proposition, must be false; and the original proposition must be true."

In the "reductio ad absurdum" the principle of Contradiction operates in connection with a course of reasoning which follows the general law of reason and consequent. In another mode of inference, which has been called "contraposition" the principle of Contradiction works alone. Let one of two predicates set forth the positive conception of a thing and the other the corresponding negative conception; of course, then, the two propositions which apply these predicates to the same subject are immediately contradictory; for example, "the man is guilty," and "the man is innocent." If, now, either of these propositions is affirmed the other may be denied, and if either be denied the other may be affirmed. Accordingly we say:

    The man is guilty; therefore,
    The man is not innocent:—
    Every righteous man is happy; therefore,
    No righteous man is unhappy:—
    Some possible cases are not probable; therefore,
    Some possible cases are improbable.

In this mode of inference the conclusion sets a negative predicate over against the positive predicate of the premise, or a positive over against the negative; and also opposes negation to affirmation, or affirmation to negation. Hence the name, "contra position."

11. The last of those three principles mentioned for present discussion is the law of the Excluded Middle, or of the Excluded Third. This law is inferior to the other two in frequency of use and in practical importance; yet is logically prior to them both, but especially to the law of contradiction. It declares that either the existence or the non-existence of a thing is always a reality, and that there is no middle object of belief between positive and negative fact; or rather no third object of belief at all. From this it follows that any proposition and, of course, each of two immediate contradictories must be either true or false. Then the principle of Contradiction adds that one only is true, and that the other only is false.

The law of Identity assumes a positive fact and asserts that it must remain so; or a negative fact, and asserts that it must remain so. The law of Contradiction, assuming a positive fact, denies the negative assertion opposed to it; or, assuming a negative fact, denies the positive assertion opposed to it. The law of Excluded Middle assumes neither positive nor negative fact, but only asserts that, in every case, there must be either one or other. Let some question be under investigation. Should, or should not, a protective tariff be levied? The Excluded Middle declares respecting each side of this question that it must be true or false: because there is no middle state possible either between being and non-being, or between truth and falsity; or rather no *third* alternative, of any description, besides the existence and the non-existence of things, or the truth and the falsity of propositions. Let us now find that one side (no matter which) is true. The law of Identity asserts that this opinion, if true, will remain true. Then the law of Contradiction adds that this side only is true, and that the other, alone, is false. And, once more, the law of Identity authorizes us to hold all that we have thus ascertained; at least till we discover ourselves to have been mistaken.

The law of Excluded Middle is sometimes stated objectively by saying, "A thing must either be or not be," and subjectively, by saying, "Every assertion must be either true or false." These formulas, however, express the law only when they are taken in a weak sense. As the statement, "The man is either a knave or a fool" may signify merely that the man has one of these characters at least — not that he has one only; so the statement, "A thing must either be or not be," might mean merely that every fact is either positive or negative — not that it may not be both at once. This last point, however, is included in the ordinary and stronger sense of the above formulas. As the statement, "The man is either guilty or innocent" does not mean that he is either guilty or innocent, or both, but that, if he is not either of these two things, he is the other; and that if he is either of them, he is not the other; so the assertion, "A thing must either be or not be," naturally signifies that one of these alternatives must be true — true only — if the other is false, and that one must be false — false only — if the other is true. This formula, therefore, unites the laws of Excluded Middle and of Contradiction in one compound law.

This combination is sometimes called the law of Contradiction, sometimes the law of the Excluded Middle, and sometimes the principle of Contradiction and of the Excluded Middle. It is really the principle of Contradiction with that of the Excluded Middle prefixed to it. That the law of Excluded Middle is of a subordinate character is evident from the fact that it is practically important only in this combination with the law of Contradiction, and as the basis for the operation of that law.

CHAPTER XV.

HOMOLOGIC INFERENCE.

1. Proceeds on one principle. 2. Widens the operation of the law of Reason and Consequent. 3. Is based upon the law of Conditions. 4. Has three modes, (*a*) the paradigmatic, (*b*) the principiative, (*c*) the applicative. 5. The common doctrine as to "deduction" and "induction." 6. The homologic principle (*a*) abbreviates ratiocination, (*b*) justifies the inference of specific effects or causes, (*c*) enables us to "reason in the general."

1. ORTHOLOGIC inference accords with, and is supported by, many fundamental laws of existence and of thought. Homologic inference follows but one such principle, namely, that *similar antecedents are accompanied by similar consequents.*

This principle assumes that logical sequences depend, not on all the circumstances which a case may present, nor even on all those included in the antecedent, or reason, but only on certain essential conditions, which together constitute an exact antecedent. When we say that a cube of wood with a base two inches square must be eight times as large as a cube of gold on a base one inch square, this consequence is seen to depend on the geometrical nature and relations of the things mentioned; and is only accidentally connected with the color, the weight, the chemical constitution, the physical properties, and the commercial value, of the cubes compared. Hence, in ordinary inference, though the antecedent is conceived of as including more than the necessitating conditions of the consequent, it is never conceived of as including all the circumstances perceivable in the case; many of these are neglected as non-essential. Our thought may even be so specially directed to the points on which the sequence depends, that it may be confined to a consideration of these points alone. Therefore inference, even when it may take place without generalization, commonly involves more or less precision and abstraction in the perception of the antecedent.

2. The homologic principle asserts that, *whenever given circumstances contain an antecedent similar to one already found to have a certain consequent, we may infer a similar consequent in connection with the similar antecedent.* This law resembles that according to which like causes are inferred from like effects and like effects from like causes; but it is much more comprehensive, because it relates to every ground of logical connection. A conclusion based on it is said to follow by "parity of reasoning"; and claims the same degree of confidence with the prior conclusion, provided the antecedent on which it depends, is precisely similar to the antecedent of the prior conclusion. This exact similarity is what is meant by "logical identity," and is often expressed by saying that the reason for the second conclusion is "the same" as that for the first.

3. The homologic principle, like that of inference in general, is closely related to the law of Conditions. It is based on the ontological law that like entities are controlled by like conditions; and this is an essential, though a subordinate, part of the general law of Conditions. Hence, too, in accordance with its origin, the homologic principle is a kind of attachment, which works in connection with the principle of Reason and Consequent; and which applies equally to every mode and degree of inference. Whether a sequence be apodeictic or problematic, actualistic or hypothetical, a similar consequent may always be inferred from a similar antecedent.

4. While homologic inference, unlike orthologic, obeys only one law, it assumes three different forms, or modes, according to the development of thought and perception in conjunction with which it is experienced. For either we may immediately infer one individual sequence from another; or we may infer general principles from individual sequences; or we may infer individual or particular sequences from general principles.

These three modes of inference may be distinguished as the *paradigmatic*, the *principiative*, and the *applicative*.

The first is named paradigmatic inference, or paradigmatization, because it is immediately founded on the use of an example (παράδειγμα), or individual instance parallel to the case in question. Aristotle mentions this mode of inference, but teaches that we first infer a general principle from the instance, or instances, given, and then, in turn, infer the individual or particular conclusion from that. Such a process, however, is not necessary, and does not always take place. The perception that one fact is logically followed by another involves, as we have said, some abstraction and precision in determining the antecedent, and more or less rejection of non-essential circumstances; and this abstraction often results in the formation of a rule of judgment: but we can reason to a parallel case without any such rule. If only we perceive that given circumstances contain a new antecedent similar to that already observed, we may immediately infer a similar consequent. The child who has enjoyed the sweetness of one lump of sugar, cries for another lump, not because of the general truth that sugar is sweet, but because he expects the second lump to affect him in the same way as the first. And the mathematician, who has demonstrated, orthologically, that the sum of the angles of the plane triangle A is equal to two right angles, immediately infers, homologically, that the sum of the angles of B, another plane triangle, are also equal to two right angles.

The second mode of homologic inference is equally dependent with the first on the law that like consequents follow like antecedents; yet perhaps not so evidently. In principiation the terms of a sequence, after being conceived precisely, or abstractly, are divested of their individuality; and thus yield a general rule, or principle. This rule is valid only because any antecedent to which it may apply, must be like the first found antecedent, and must, therefore, have a consequent similar to the first consequent.

Principiation is the generalization of a sequence. It is more than the generalization of thought; inasmuch as the forms of thought produced by it are accompanied with conviction. Neither can this process be adequately designated by the term "induction." Induction is only that species of principiation by which the laws, or general causational sequences, of Nature, are determined. Any general truth whatever — for example, any axiom or postulate of mathematics or of metaphysics — may be obtained by principiation. This, indeed, is the only way in which axiomatic truth is originally obtained. The doctrine that all general principles, or rules of reasoning, are derived by principiation from perceptions of individual connections, or sequences, is the first and most fundamental principle of philosophical method.

Principiated truth is chiefly valuable because it may be stored up in the mind as a basis for future inference. For whenever afterwards a case arises such as a general principle contemplates, we can infer a consequent such as that principle requires. And this inference we style "applicative," because it consists in the application of the general truth to the particular case. It evidently depends wholly on the homologic law.

5. Most logicians distinguish this applicative inference as deduction, because it is the "bringing down" of a general principle to a specific case. This use of language need not be rejected; though the word "deduction" may signify "bringing from" as well as "bringing down," and often indicates any kind of formal inference. But a serious error is inculcated when *deduction*, — that is, applicative inference — is contrasted with *induction*, and we are taught that all inference belongs to one or other of these two modes. Deduction, or application, should be contrasted with principiation, of which induction is only an important species; and then even principiation and deduction, so far from being the only modes of inference, are merely the more formal modes of homologic inference.

6. Regarding, now, this kind of inference in the general, let us note that our use of it results in *three practical benefits*. For, in the first place, the homologic law abbreviates reasoning. A mathematician, having discovered, by a course of demonstration, that the solidity of a cone is measured by one third the product of its base and altitude, immediately employs this method of calculation for another cone, and for all cones. The solution of the individual problem is accepted as the solution of others exactly similar. It originates a general

truth, a law of inference. Without the homologic principle we might conceive of such a law, but we would have no ground to believe that it expressed truth.

In the next place, the homologic principle enables us to foretell natural consequences, and to ascribe effects to their proper causes. In such judgments, we do not simply substitute a shorter process for a longer one; we form specific inferences which could not be formed in any other way. Those judgments and reasonings which are based purely on the necessary nature of things may take place orthologically, and without reference to any previously perceived case of similar connection: the homologic principle may be dispensed with in such reasonings; it only renders them shorter and easier, as in the case of mathematical calculations and demonstrations. But inferential judgments concerning specific causational relations must rest on previously perceived cases of similar connection, or consequence; they must be formed homologically. For the peculiarities of specific causes and effects are perceived only as belonging to the actual constitution of the Universe — not as belonging to the necessary nature of things. These peculiarities appear to have been ordained by the power which first created and constituted the Universe and its component parts. They become known to us only by actual observation, or experience. We can, without reference to previous experience, say that every change or beginning of existence must have some cause, and that similar powers under similar conditions will produce similar results; but we cannot tell, except from a previously observed case, that a specific causational antecedent, and a specific causational consequent, will accompany each other.

Inductive principiation, therefore, differs from axiomatic, and inductive reasoning in general from that which is mathematical or metaphysical (or ontological), in that the former is necessarily founded on observation; which is not the case with the latter.

Thirdly, and finally, the homologic principle justifies reasoning in the general, so that a process of argument may be conducted throughout, and its conclusion given, *in general terms;* after which, of course, the conclusion may be applied to any individual case, or cases. We have seen how every individual inference may, through principiation, yield a general inference, or law of reasoning. Should there now be a series of such principles, or generalized inferences, so related to each other that the antecedent of the second is the consequent of the first, the antecedent of the third the consequent of the second, the antecedent of the fourth the consequent of the third, and so on to the end of the series, it is clear that the last consequent may be inferred from the first antecedent. Because the antecedent of an antecedent is the antecedent of the consequent also. And evidently that same homologic principle which justifies the formation of general inferential propositions, also renders it possible for us thus to reason consecutively by means of them.

The great merit of the Aristotelian doctrine of the syllogism is that it sets forth the laws and forms of the correct sequence of generalized inferences; and so supplies a test of all reasoning. For every step in a course of reasoning, except the application of the conclusion to some individual case, or cases, may be conceived and expressed in the general.

CHAPTER XVI.

INDUCTIVE REASONING.

1. Induction, the principiation of causational sequence. 2. Often signifies, not this act, but a process; 3. In which there are five stages: (*a*) observation, (*b*) supposition, (*c*) principiation, (*d*) criticism and suggestion, (*e*) deduction. 4. Observation includes experiment. 5. Inductional supposition is an homologic suggestion, and involves more than the association of ideas. 6. Inductive principiation is the essential part of the process. 7. Scientific criticism argues from (*a*) the ontological law of causation, (*b*) the ascertained constitution of the universe. 8. Inductional deduction. 9. The maxims of scientific suggestion imply that Nature has "an intellectual constitution." 10. They assert that Nature (*a*) has a fixedness of operation, (*b*) abounds in analogies, (*c*) uses reliable signs, (*d*) is parsimonious of instrumentalities, (*e*) is simple in her methods, and (*f*) is governed by design, or Final Cause. 11. Ontological principles determine (*a*) the method of agreement, (*b*) the method of difference, (*c*) the indirect, or analogical, method of difference, (*d*) the method of residues, and (*e*) the method of concomitant variations.

1. THE word "induction" primarily signifies that act of principiation in which some law of causational sequence is inferred from the perception of some individual sequence, or sequences. This act presupposes causational antecedents and consequents. A causational antecedent is any definite combination of circumstances which has and exercises the power to produce a given consequent. It therefore always includes, or implies, an efficient agent and the conditions of its operation, or the conjunction of such agents and their operations; but it may also include other elements in union with these. The efficient agent, or set of agents, and the conditions of its operation, are the exact philosophical cause of an effect, and in a manner reciprocate with the effect; for they may always be inferred from the effect. Thus the rising of the sun is the exact, or reciprocating, cause of day; and a certain combination of

sodium and chlorine is the exact cause of salt. In such cases we may infer effect from cause, and cause from effect. But an ordinary cause contains the exact cause within some envelopment or other; so that the same effect is often said to result now from this cause and now from that one. The burning of fuel is a cause of heat, but it is only one out of a number of causes; and extreme disease is only one particular, or specific, cause of death. To determine in any case what should be regarded as the cause, or causal antecedent, of a given effect is not always easy when we are only seeking to define some specific cause; and it is often very difficult when we would clearly discern a reciprocating cause. But, in either case, after that determination, the act of induction which follows is perfectly simple. It is merely a generalization based on the homologic principle.

2. Frequently, however, because of the importance of the principiative act, the word "induction" is used comprehensively, and signifies a process in which principiation is only an essential part; so that commonly, when we say that a principle has been ascertained by induction, or by inductive reasoning, we mean that some law has been determined by a process which has terminated in principiation. Sometimes, even, we speak of an individual or particular conclusion being reached inductively, because we have come to it through a process in which we first gain a general principle and then apply that principle.

3. This process of inductive reasoning varies in the extent and variety of its parts, according to the requirements of each case; but in its fullest development may be divided into five parts, or stages. *First*, there is a careful observation and a "simple enumeration" of those facts, or phenomena, which appear to contain "instances" of the sequence to be investigated; *secondly*, there is a more or less definite apprehension or conception of the sequence in the individual cases, following upon an analysis of each instance, things evidently non-essential being rejected; *thirdly*, there is the act of induction, or principiation; *fourthly*, a critical testing and elaboration of the law, whereby our conception of it is rendered more ade-

quate and truthful; and whereby also we may be prepared for the apprehension of some higher law; and *fifthly*, there may be a deduction from the law in combining it with other laws already ascertained, or in applying it to individual cases.

4. The first of these stages, while presenting no theoretical difficulty, demands diligence and skill from the investigator. At least, great pains are necessary if we would ascertain the less patent laws of the Universe. Long journeys, costly instruments, accurate records, the watching and waiting of years, may be called for. Moreover, Nature must often be made to work under conditions furnished by the student, in order that the results of experiment may be added to those of simple observation.

5. The second step in inductive reasoning — that is, our first formal conception of the sequence — involves some power of penetrative and constructive judgment. It would be impossible if the human mind could not often perceive causational sequences, as such; or if, in cases of question, we could not form a more or less probable supposition, or hypothesis, regarding the character of a cause, or of an effect. Inasmuch as we do not directly observe any force, or efficiency, producing the changes which occur around us, but only a succession of phenomena, or events, some philosophers have denied that the relation of cause and effect is anything more than uniformity of succession; they have taught that our apparent perception of power, or force, is either a delusion of the mind, or, at the most, a form of thought which the mind imposes on phenomena, and which has no objective significance, although, perhaps, it may represent a unity — or a strong association — of ideas. Such teachings are unsatisfactory; instead of explaining, they explain away what every human being naturally and necessarily believes. We prefer that doctrine which asserts that *all man's knowledge of the causal relation, and of specific causes and effects, originates in his perception of those changes which take place within, or in immediate connection with, his own body and his own soul.*

The various powers and operations of spirit are seen through self-consciousness; while the essential attributes of matter and the specific qualities of material substances, become known to us as related to our own bodily efforts and sensations. Solidity, or the space-filling quality, is first perceived as belonging to the members of our own bodies, and is then inferentially assigned to things about us. This is the case also with that force, or power of propulsion, which shows itself in muscular exertion and resistance. The sense-affecting qualities of objects are powers residing in them; and which we ascribe to them because we find them to exert these powers upon us on the recurrence of the proper conditions.

The full discussion of the law of cause and effect, and of specific causational perceptions, belongs to metaphysical psychology. What has now been said may indicate in what way the mind becomes qualified to distinguish between a true causational sequence and an accidental succession of events. The power to recognize causes is a logical outgrowth of man's original and immediate perceptions; nor is there any rule whereby causal may be surely distinguished from merely temporal antecedents except that a cause always involves efficiency, and that our recognition of any kind of efficiency must be founded on a first knowledge gained in man's personal experience.

Those suppositions, or hypotheses, which we form when only some conditions of a causational sequence can be seen and others must be conjectured, yet more evidently than our unquestioned perceptions, are based on the knowledge which we already possess of causes and causal laws. They are not free imaginings; they are conceptions of antecedents in which causes and conditions more or less similar to others already known, are so combined that they may be supposed capable of accomplishing given results. This is the origin both of theories by which phenomena are explained, and of practical inventions by which phenomena are produced. Therefore, also, no man is properly qualified to make discoveries or inventions who has not mastered all the knowledge which bears in any way on the field of his investigations.

6. When a causational sequence is obviously and exactly perceived by the observant student, as, for example, often hap-

pens in decisive chemical experiments, the process of inquiry may be made to close with that generalization, or principiation, which has been mentioned as the third step in inductive reasoning.

7. But if the student has only formed a probable or incomplete hypothesis, a fourth stage of investigation is necessary in order to remove the doubt, or to remedy the imperfection: the judgment, or hypothesis, which has been formed must be subjected to a process of trial and amendment. This process is not of the nature of principiation; neither is it essentially deductive, though it may be regulated by rules. It consists of a further questioning of instances and experiments, both new and old, together with a more methodical interpretation of them according to those relations by which causes are perceived to be conditioned. For, in our cognition of causes and effects, we intuitively perceive such things as the following to be necessary, viz., that, in the absence of any cause there is no change, so that things remain as they are — that every change has an adequate cause, or a variety of adequate causes, so that, if the effect take place, some adequate cause may be inferred; but if the effect do not take place, no adequate cause exists for it — that a part of a cause may exist without any production of the effect; but that, if an effect take place, every part of the cause producing it must have existed — that a conjunction of effects and a corresponding conjunction of causes involve each other — and that the same cause (that is, the same, or a precisely similar, potency, or combination of potencies, under the same combination of conditions) produces the same effect. These judgments are *intuitive perceptions of the direct, and of the corollary, workings of the great ontological law of causation.* We make them constantly in cases which frequently occur, and finally, by principiation, derive from them those rules which are the fundamental canons of inductive — or inductional — criticism.

8. After a law of causation has been determined, either directly or after critical elaboration, it may be applied to the inference of individual consequents. The result so obtained, because it is the ending of an inductive process, is sometimes

said to be reached inductively. And, with more reason, perhaps, the same language is employed when two or more laws of causation are combined so as to form a new compound law. For, though this also is a case of deduction, it is doubly the result of inductive reasoning. Important inquiries have been answered in this way; and this is likely to occur more frequently as the knowledge of principles increases. Hence many think that deduction will hereafter share, equally with principiation, in the honors of scientific progress.

---

9. We have now briefly sketched those mental operations whereby we arrive at conclusions regarding causational sequences, and which are often grouped under the head of "Inductive Reasoning." If our analysis of these operations be correct, it will prepare us to understand philosophically a certain set of maxims which have always guided scientific conjecture, and a certain set of rules whereby scientific theories are often tested.

The maxims to which we refer are all connected with a belief, universally diffused among men, that Nature, or the Universe, has an orderly, and, if we may so speak, an intellectual, constitution. We do not mean by this that Nature possesses any power of thinking, but only that the Universe, in all its departments, is evidently the production of rational plan and purpose; and therefore, also, is such as rational thought can understand and appreciate. Some explain this conviction as an immediate intuition of the mind; it seems nothing more than an homological inference from the formation and use of plans and instrumentalities by man himself. Observation shows that intelligence is the only knowable cause for any continued and complicated adjustment of means to desirable ends; and reflection on the nature of things convinces us that no other conceivable cause can adequately account for such an adjustment. Therefore, discovering wise adaptations, first in one natural arrangement and then in another, we spontaneously conclude that rational methods pervade every part of the Universe. Moreover, men become greatly confirmed in this conclusion as they progress in their knowledge of the works of Nature.

Such is the origin of those directive maxims which presuppose the intellectuality of the Universe. They are not self-evident truths, but the results of observation and thought; and they may be regarded as fundamental parts of that prior knowledge which qualifies one for the second stage of the inductive process — that is, for the true apprehension of a sequence, or for wise conjecture concerning it.

10. The most common of these maxims asserts that *the course of Nature is fixed and uniform*. By this we are not to understand that the arrangements of the Universe are absolutely unchangeable, but only that they have a permanence which characterizes every wisely formed constitution of things; nor are we to understand that Nature is wanting in variety; for her variety is multitudinous; but only that lines of law and order are traceable in every department of the Creation. The Universe, animate, or inanimate, organic or inorganic, is composed of genera and species of things. Each of these conforms to a certain type and has its own method of existence; and can, therefore, be rationally comprehended. Hence the different branches of scientific knowledge correspond to different systems of permanent uniformities.

The maxim that *Nature abounds in analogies* is little else than a corollary of that just considered. When we prefer one theory to another because it accords better with the analogy of Nature, we simply recognize the intellectual unity and stability of the Universe. For, as a matter of fact, Nature is found to use similar methods to effect similar ends, even though quite other methods might have been employed. A notable instance of this is the radical similarity in bodily structure of all the larger animals, whether beasts, birds, or fishes, even while the greatest dissimilarities arise in accordance with the necessities of their different spheres of life.

In the next place, it is constantly assumed by scientific men that *Nature uses reliable signs to indicate her agencies*. There might be a universe in which like causes, or agencies, would always produce like effects, but in which, nevertheless, we could not be confident that any agency which seemed to us of a given kind was really so. But now, in the constitutions of things, immediately perceptible qualities have been so united to other qualities that they may be taken as indications of the entire natures to which they belong. Every kind of metal has a color and a specific gravity which mark that metal only; and which suggest and represent to us the whole complex of its qualities. The appearance of any animal, or insect, or plant, of any fruit or seed, brings before us the complete natural history of one specific organism. In short, Nature takes pains, not only that her methods should be fixed and orderly, but also that they should be easily apprehended by beings of a finite intelligence. Resting on fixed and observable signs, and exercising proper diligence, man obtains a usable knowledge of causes and becomes qualified for the control and management of natural agencies.

Again, superintending wisdom is recognized in the maxim that Nature, though lavish of her expedients, *is parsimonious of her instrumentalities*. She accomplishes an immense variety of results with the smallest possible variety of agents. However peculiar a problem may be, no new agency is introduced unless it be necessary; there is rather some extraordinary modification of an ordinary agency; as may be seen in the trunk of the elephant, the tail of the otter, and the wings of the flying fish. A perception of this parsimony in the use of powers and instruments gave rise to the adage "Entia non sunt multiplicanda praeter necessitatem." If phenomena can be explained as well by supposing one kind of agent as by supposing two, or as well by supposing two as by supposing three, the preference is to be given to the smaller number. Because differences of specific gravity account for the heaviness of some bodies and the lightness of others, only one agency is recognized in both these phenomena. The law of gravitation is found sufficiently to account for the continued motion of the heavenly bodies; and therefore we reject the supposition of any peculiar celestial force. In like manner, the hypothesis that, in the successive stages of creation, certain organic forms were built upon others and immediately produced by giving to a departing species the power to produce a successor better qualified for life under new conditions, cannot be condemned as unphilosophical, unless it should be found to conflict with fact. It may relate, however, not to the workings of Nature, but of the power that produced Nature.

Closely allied to the law of parsimony, and perhaps radically identical with it, is the maxim that *Nature is simple in her methods*. Scientific men always prefer the simpler explanation, provided that, in other respects, it is equally satisfactory with the more complex. But this simplicity of Nature is to be understood in a relative rather than in an absolute sense. Some of her arrangements are complicated, and resemble very ingeniously constructed instruments or machines. This is always the case when the work to be done includes a large variety of movements or functions, such as are provided for in the mechanism of the human arm, or eye, or of man's body as a whole. Yet, however complicated a natural organism may be, the thoughtful student is amazed both at the simplicity of its several contrivances, and at the neatness with which they are united in one effective arrangement. No part of the system is superfluous, or out of place.

Finally, the intellectuality of the Universe is expressly asserted in the doctrine, that *Nature is governed by final causes*, or by intelligent design—that wisdom operates in the Universe through means adapted to the accomplishment of ends. This doctrine has always influenced speculation; and has always been a teaching of philosophy. Nor need we wonder at this, since the doctrine only expresses a natural and rational judgment. The Stoic aphorism, "God and Nature do nothing in vain" (ὁ θεὸς καὶ ἡ φύσις οὐδὲν μάτην ποιοῦσιν), and Aristotle's conception of the final cause (τὸ οὗ ἕνεκα), simply formulate a general conviction of mankind in regard to the origin of the phenomena of the Universe. For the Peripatetic division of causes, or rather of causal conditions, into the material, the formal, the efficient, and the final, is not really a theory of causation in the abstract, but a cosmogony. It analyzes the causal antecedent of the Universe into four constituents—one of these being design.

Moreover, it is worthy of remark that Aristotle did not consider the world to be itself capable of thinking or deliberation; for, he says, that would be "as if the art of ship-building were in the timber," or as if any machine had the intelligence to construct itself. Indeed, the fact that Nature, notwithstanding her wonderful excellence, sometimes produces abortions and monstrosities, indicates an imperfection which probably is inherent in every created agency. In herself Nature is only a marvellous system of powers and laws which operates throughout the Universe, and which, though unintelligent, may be termed intellectual, because it is the production and the reflection of creative thought.

By some philosophers inquiry after final causes has been condemned as fruitless. This objection applies only to cases in which conjectures are made without adequate support in existing analogies, and are rested upon as probable without experimental evidence. Mere theorizing respecting the work for which some arrangement or agency is designed, when separated from the observation and investigation of facts, has originated many strange explanations of natural phenomena; and is worse than fruitless. But hypotheses formed after the analogy of known adaptations, and followed by investigation, have often led to the discovery of truth. Harvey, observing valves in the veins and in the heart, first conjectured, and then discovered, the circulation of the blood. Physiologists discuss every bodily part in the light of some end for which they suppose it to be intended; and they declare that every part is an organ, with a function of its own.

11. Let us now glance at those canons of experimental enquiry, whereby hypotheses are tested, and which are used chiefly in the fourth, or critical, stage of inductive reasoning. For rules are not needed when every causal condition of a sequence is clearly perceptible, but only when the exact nature of the cause is in doubt. This is especially the case when the cause of some effect is involved in a confusing complex of circumstances; then a work of determination and of elimination becomes necessary. A less or a greater number of directions may be given for this work according to the com-

prehensiveness of each rule, but the following five canons discussed by Mr. J. S. Mill, under the head of "methods of induction," are certainly such as every careful thinker must use. They are all outgrowths of the radical law of causational sequence.

The first rule is that which governs the "*method of agreement.*" When two or more cases of sequence, which have the same consequent, have only one circumstance, or set of circumstances, in common, the antecedent of the consequent is to be sought for in their common part. If a certain fever prevail in two or more localities, in both of which the air is tainted from decaying vegetation, but which differ in all other respects, we say that malaria is the cause, or an essential part of the cause, of the fever. If cucumbers thrive whenever they are planted in rich mellow earth, and enjoy an abundance of warmth, light, and moisture, and if they call for these conditions only, we say that we have found the right way for the cultivation of cucumbers.

The second rule controls the "*method of difference.*" If various cases which produce a sequence differ, severally, from other cases which do not produce it, only in the presence of a certain antecedent which is uniformly absent when the sequence is absent, that antecedent is, wholly, or partly, the cause of the sequence. If, on the other hand, a supposed cause be found present in cases where the sequence does not occur, as well as in cases in which it does occur, it cannot be a true and sufficient cause. Since dew falls always on clear nights, but never when the sky is clouded, we ascribe the formation of dew to the cooling of the surface of the earth by radiation. Since all living things breathe the air, and cease to live when prevented from breathing it, we say that air is essential to animal life. The method of difference presupposes the method of agreement, and is built on it. It is applicable whenever a given consequent fails to occur, and this failure is either in accordance with our expectation or in opposition to it. If the failure take place in accordance with our expectation and along with the absence of the supposed antecedent, our theory is confirmed; but if it fail in opposition to our expectation and notwithstanding the occurrence of the supposed antecedent, our theory must be rejected. If a certain compound, expected to explode on ignition, will not explode, our conception of the antecedent is evidently wrong. After learning this, if we still desire to find a new explosive mixture, we must amend our hypothesis, and renew our experiments and our examination of instances.

Sometimes a single instance of a sequence, being distinct and free from all complication, is sufficient to determine a law. Yet oftener a pair of experiments, or observations, one using an antecedent and the other leaving it out, are sufficient. In such cases we can scarcely be said to need or to follow either the method of agreement or that of difference; we simply decide at once according to the principles of the law of causation. But when elimination and determination are necessary, we are greatly helped by analyzing a number of instances.

The third rule sets forth *the indirect method of difference*. Sometimes no cases of the non-occurrence of a consequent can be found which differ from cases of its occurrence merely in the absence of some antecedent. If then we only can find cases of the non-occurrence, which are more or less similar to the cases of the occurrence except as to the presence of any similar antecedent, we may consider that antecedent to be wholly, or partly, the true cause. No species of quadruped, or other animal that is warm-blooded, differs from the ordinary quadruped, or other animal, in being cold-blooded. But we can find animals that are cold-blooded, and we may reason from their constitution by a kind of negative analogy. Thus, says Mr. Mill, "If it be true that all animals which have a well-developed respiratory system, and therefore aerate the blood perfectly, agree in being warm-blooded, while those whose respiratory system is imperfect do not maintain a temperature much exceeding that of the surrounding medium, we may argue from this two-fold experience, that the change which takes place in the blood by respiration, is the cause of animal heat." This third method is simply a special form of the method of difference; and is guided by a reference to the analogies of Nature.

The fourth rule presents what Mr. Mill calls the "*method of residues.*" If we subduct from any complex of phenomena such parts as are known to be the effects of certain antecedents, the cause of the residual phenomenon, or phenomena, is to be found in the residue of the antecedents. This method endeavors to isolate a case mentally which cannot be isolated in fact. The principle of it is that by which we find the weight of a load of hay in subtracting the weight of the wagon from that of the wagon and the load. But by the observation of residues we determine separate kinds of causes or of operations, as well as the respective shares which two or more increments of the same cause may have in producing a result. Newton, wishing to know how far an ivory ball suspended by a cord and allowed to strike a hard surface, would rebound by the force of its own elasticity, first of all caused it to swing freely in the air, and measured the loss of motion produced by the resistance of the air during each vibration. Then adding to the length of the rebound the loss of distance incurred in the half-vibration of equal length, he obtained the entire effect of the elasticity. The observation by astronomers that the planet Uranus was sometimes retarded and sometimes accelerated in its orbital course, so as not to be in its calculated positions, led to the discovery of the planet Neptune, as the cause of the aberrations. So also the fact that comets generally do not return from their distant journeys till after the expiration of the predicted time, has suggested the existence of some cosmic ether, or other medium, capable of obstructing the motion of such bodies.

The fifth rule explains the "*method of concomitant variations.*" If a phenomenon which is either continuous or recurrent, varies in a manner to correspond with the variations of another phenomenon, these phenomena are connected through some law of causational sequence. The mere concomitance of the variations does not indicate the specific mode in which the phenomena are related to each other. It does not, for example, show which is cause and which effect, or whether both are effects of the same cause; but the nature of the specific relation is commonly easily determined. When quicksilver was observed to expand in proportion to the heat about it, no one hesitated to believe that heat is the cause of the expansion. So friction is proved to be the cause of heat, when it is found that heat is evolved exactly in proportion to the amount of force expended in rubbing one substance against another.

The law of concomitant variations is a specific application of the principle that every cause and its effect mutually correspond — the presence or absence of the one involving the presence or absence of the other. But it enables us to interpret a peculiar class of cases, in which the cause never ceases from operation; and in which, therefore, the ordinary method of difference is not available. The fact that the tides follow the moon, and that the high tides attend the conjunction of sun and moon, indicates that the rising and falling of the ocean results from the attraction of these bodies. The seasons evidently result from the sun's changes in latitude. A correspondence in the periodical prevalence of "magnetic storms," of the Aurora Borealis, and of solar spots, with certain recurrent positions of the planets Jupiter, Saturn, Venus and Mars, has led some to think that these planets are the prime movers in a remarkable set of meteoric phenomena.

CHAPTER XVII.

HYPOTHETICAL AND DISJUNCTIVE REASONINGS.

1. Inference is also actualistic or hypothetical. 2. The so-called hypothetical syllogism is translative. 3. The law of logical transfer. 4. Translative inference is either express or implicit. 5. The simple hypothetical, or translative, syllogism has two modes: (*a*) the *ponendo ponens*, (*b*) the *tollendo tollens*. Both explained. 6. Logical disjunction is a complicated style of hypothetical inference founded on either (*a*) contrariety or (*b*) contradiction. 7. Contrariety explained. 8. It is the ground of the weak disjunctive syllogism; which has one mode, the *ponendo tollens*. 9. Contradiction is either categorical or consequential. 10. Two contraries become contradictories when the non-reality of either involves the reality of the other. 11. Only a pair, not a series, of things can be mutually contradictory. 12. The strong disjunctive syllogism has two modes, the *ponendo tollens* and the *tollendo ponens*. 13. The dilemma is an hypothetical syllogism, with a plural "major" and a disjunctive "minor." It is either (*a*) constructive or destructive, (*b*) simple or complex.

1. WITH reference to the mode of its sequence, inference is either apodeictic or problematic; with reference to its dependence on previous perceptions of logical connection, it is either orthologic or homologic; and with reference to the character of the conviction produced, it is either actualistic or hypothetical. Actualistic inference is founded on what is known or believed to be fact; and its consequent is accepted as fact, either absolutely or possibly or probably, according to the modality of the sequence. Hypothetical inference rests on mere supposition, and asserts only what *would* certainly or possibly or probably be fact provided the antecedent were a reality.

Every hypothetical proposition is illative, or inferential, in its nature. This is especially evident in the case of fully expressed hypotheticals. "If chlorine be a gas, it is elastic," asserts that a certain consequent must be true if a certain antecedent be true. The only difference between an hypothetical proposition and an hypothetical inference is that the former emphasizes the consequent rather than the antecedent; while the inference dwells equally on both.

2. That form of reasoning, however, which logicians style the "hypothetical syllogism," should not be confounded with mere hypothetical inference. It is really an hypothetical inference *with an addition which has the effect of depriving the process as a whole of its hypothetical character.* When the statement, "If chlorine be a gas, it is elastic," is followed by the assertion, "chlorine is a gas," the object of this addition is to assert the reality of the antecedent, and thereby to change the character of the inference from hypothetical to actualistic. This appears in the conclusion, when we assert, for a fact, that "chlorine is elastic."

In consequence of the application of the term "hypothetical" to syllogisms of this kind, some ambiguity arises when this adjective is used with reference to inferences generally. Were a special name desired for inferences and arguments purely hypothetical and unchanged by actualistic addition, they might be distinguished as *suppositive*. The following would be suppositive inferences: "If air be a substance, then it occupies space; if trees spring from seeds, then these trees do so; if all gases are elastic, and oxygen is a gas, then oxygen is elastic." These inferences would become "hypothetical" syllogisms, if additions were made to them asserting that their premises set forth reality.

3. The law according to which an hypothetical is changed into an actualistic inference is a very simple one, and may be considered a corollary, or supplementary part, of the general law of antecedent and consequent. It recognizes the difference between two radical modes of conviction, and operates whenever we apply hypothetical statement to actual fact. Asserting the reality of the antecedent, it claims reality for the consequent. This law might be styled the principle of logical transfer, becauses it enables us to transfer an assertion from one kind of conviction to another; and syllogisms whose antecedents are constructed in accordance with this principle, might be called translative reasonings, or inferences.

4. The working of this law, may be either express or implicit. Its express operation occurs when the minor premise, as it is called, asserts fact immediately and exclusively. This takes place in all translative reasonings concerning existing individuals; as, for example, in the syllogism, "If Socrates be virtuous, he merits esteem; he is virtuous; therefore he merits esteem." The implicit working of the law appears when the reasoning immediately concerns general objects, or logical classes. In saying, "If oxygen be a gas, it is elastic: oxygen is a gas; therefore it is elastic," the minor premise has an actualistic force; yet not simply and directly, but only as implicated with a general truth. In other words, the assertion, "oxygen is a gas," has a double significance; first, it presents a principle which applies, not only to existing oxygen, but to any that ever may exist or may have existed; and secondly, it contains the implication that some oxygen actually exists and is a gas. The "hypothetical," or translative, syllogism depends on this assertion, in the second premise, of the reality of the antecedent supposed in the first premise; and only accidentally uses a general truth or principle for this purpose. The proper force of general principles in reasoning will be considered hereafter, in connection with syllogisms of another nature.

Some define the hypothetical syllogism as that mode of reasoning which is governed by the principle of antecedent and consequent; and say that other modes of reasoning follow other principles. Though this is not true, we must allow that the translative inference is specially related to the generic law of inference; inasmuch as the law of logical transfer is not only, like other principles, subordinate to the law of reason and consequent, but pertains to the operation of that law.

5. The law of reason and consequent works in two ways; we either *assert the consequent with the reason*, or we *deny the reason with the consequent*. Hence, also, the law of logical transfer has a double operation. That is to say, after an inference has been made hypothetically, we may then either assert the reason or deny the consequent actualistically, and thereupon assert the consequent or deny the reason actualistically.

Here we must determine exactly what is meant by asserting and denying; for it might be supposed that assertion always signifies the setting forth of something as existing, and denial the setting forth of something as non-existent; whereas the terms have wider meanings. Ordinarily we infer from one positive fact to another, that is, from one case of existence to another. But, in addition to this, we infer from existence to non-existence, from non-existence to existence, and from non-existence to non-existence. There are, therefore, four styles of inference; which may be illustrated, as follows: "If the man has consumption, he will soon die," (from existence to existence); "if the formation be granite, it does not contain coal," (from existence to non-existence); "if there be no food, we must suffer hunger," (from non-existence to existence); "if there be no fuel, there can be no fire," (from non-existence to non-existence). Now to assert the antecedent or consequent in any of these inferences is to present it as a reality, whether it be a fact of existence or a fact of non-existence; and to deny the consequent or the antecedent is to deny its reality, whether that be the denial of existence or of non-existence.

In order to express technically those wide conceptions which we have now explained, logicians sometimes call the assertion of the antecedent, and that of the consequent which it involves, the "positing," or "placing," of a statement of fact; and they have termed the denial of the consequent, as well as that of the antecedent, the "sublation," or taking away, of a statement of fact. They also name that form of inference which depends on the "placing" of the antecedent the "*modus ponendo ponens*," or more simply, the "*modus ponens*"; and that which follows the sublation of the consequent the "*modus tollendo tollens*," or more simply, the "*modus tollens.*"

This phraseology has the additional advantage of indicating that antecedents arise from assertion and denial, not simply because something is asserted or denied, but because, also, there is a presupposed subject, or case, or set of circumstances, in relation to which the positing or sublation takes place.

Moreover, as according to the law of contradiction the denial of existence involves the assertion of non-existence, and the denial of non-existence the assertion of existence; instead of merely denying the consequent, we may, and often do, assert its contradictory; and thereupon deny the antecedent. Hence the negative part of the law of logical transfer may assume the form, "contradict, or assume the contradictory of, the consequent, and you may deny, or contradict, the antecedent."

6. We have now considered those simple and primary modes of "hypothetical" reasoning which are expressed by the ordinary "conditional syllogism." A more complicated style of translative reasoning, which, however, is explainable on the same general principles, appears in what are called "disjunctive" reasonings.

Logical disjunction is either partial or complete. The first exists *when it is impossible that two things should be true together, so that the placing of either involves the sublation of the other.* This is the disjunction of contrariety. The second arises *when, in addition to the foregoing opposition, two things cannot be untrue together, so that the sublation of one involves the placing of the other.* This is the disjunction of contradiction. As contradiction presupposes contrariety we shall consider the latter first.

7. The nature of contrariety, and its relation to inference in general, may be understood from the fact that a case of this mode of opposition may be produced by the denial or contradiction of any consequent of necessity. To illustrate this point let us take the sequences already mentioned:

> If the man has consumption, he will die soon;
> If there be no food, we must suffer from hunger;
> If the formation be granite, there cannot be coal in it;
> If there be no fuel, there cannot be any fire.

The first two of these are sequences of positive necessity, the one with an antecedent of existence, the other with an antecedent of non-existence; the second two are inferences of negative necessity, one having a positive and the other a negative antecedent. If now we deny, or take the contradictory of, the consequent in each sequence, retaining the antecedent unchanged, we shall have the following pairs of contraries:

Consumption — continued life;
No food — no suffering from hunger;
Granite — coal in the formation;
No fuel — fire.

Assert any one of these contraries, and you must deny, or contradict, its fellow. If the man have consumption, he cannot have continued life; and if he have continued life, he cannot have consumption: if we have no food, we cannot be without suffering from hunger, but must suffer from that cause; and if there be no suffering from hunger, we cannot be without food, but must have a supply: and so on with the other contraries. In general, therefore, we say that *anything and the contradictory of any necessary consequent of it, are contraries*. The reason for this is that the consequent of a necessitating antecedent is a condition of that antecedent. Evidently, it is impossible for a thing to exist while the contradictory (or any contrary) of any of its conditions exists; or for any contradictory (or contrary) of a condition to exist while the thing conditioned exists.

The perception of contrariety, however, does not depend on a previous perception of necessary sequence; indeed, it commonly takes place independently. For two things may be directly perceived to be of such a nature that the existence of one of them conflicts with, that is, involves the non-reality of, some condition of the other; in which case they are, and must be, contraries. *This incompatibility of one thing with others is as much a part of the nature and constitution of things as the compatibility of one thing with others is, and may be as directly perceived.* For example, it is as immediately evident that two bodies cannot occupy the same space at once, and that if the one is there the other is not there, as that a body must occupy some space, or that it may occupy any sufficient space which would be otherwise unoccupied.

Contrariety is especially noticeable when a number of natures, or things predicable, which have a common character, have also such peculiarities that no two of them can belong to the same individual subject. Hence *the co-ordinate species in a correct logical division* are contraries one to another with reference to their inherence in the same subject. For instance, if an object is of any one color, say red, it cannot be of any other color at the same time; if a triangle be isosceles it cannot be equilateral or scalene, or if it be equilateral it cannot be isosceles or scalene.

8. The inference of contrariety can be expressed in the same way as "conditional," or simple hypothetical, inference, but it differs from the latter in the peculiar indirectness with which the sequence is conceived. The consequent of simple, or ordinary, hypothetical sequence is immediately conceived and asserted as true; that of contrary inference is obtained by conceiving first of something and then of the immediate, or categorical, contradictory of that something, and is the assertion of this contradictory. Conceiving of "red" as a contrary of "white," and then of its contradictory, "not red," we say that, if the paper is white, it is not red. Contrary inference, also, has a doubleness, because each contrary may be, and commonly is, conceived of as being, in its turn, antecedent to the non-reality, or to the immediate contradictory, of the other.

This doubleness may be expressed with any pair of contraries, if we follow the formulas, "A cannot be both B and C; it is B; therefore it is not C," and "it is C; therefore it is not B." "The triangle cannot be both equilateral and scalene; it is equilateral; therefore it is not scalene," or "it is scalene; therefore it is not equilateral."

Argument of this form may be distinguished as the weak disjunctive syllogism. It admits only the *"ponendo tollens."* The ordinary, or strong, disjunctive syllogism, as we shall soon see, has a *"modus ponens,"* as well as a *"modus tollens."*

9. This brings us to that thorough-going form of disjunction which is technically called "contradiction." For *contradictory opposition includes contrariety and something more.*

First, then, we say, negatively, that the disjunction of contradiction is *not at all limited to that opposition which is based on the law of contradiction and excluded middle.* This law relates to any pair of propositions which set forth the existence and the non-existence of *the very same thing;* and asserts that if one of them be true the other is false, and that if one be false the other is true. The contradictories of which we have made mention above are of this sort. But contradictory inference in general is chiefly occupied with propositions which set forth the existence or non-existence *of two different things;* and asserts the falsity of either of these propositions because of the truth of the other, or the truth of either because of the falsity of the other. To account for such inference we must assume, not merely the law of contradiction, but also an operation of the general law of antecedent and consequent additional to that which appears in immediately self-evident contradiction.

Those propositions which set forth the existence and the non-existence of the very same thing, may be styled *categorical contradictories;* for this adjective sometimes indicates that a statement is absolute, or unaccompanied by any reason. Such statements as "the man is guilty" and "the man is not guilty," are categorical contradictories; because their opposition, though founded on reason, takes place according to a law of whose operation the mind is scarcely conscious. The reason for such contradiction is considered only by logicians; it is the "law of Contradiction." But those propositions which set forth two different things or natures which conflict with each other both as to existence and as to non-existence, may be called *consequential contradictories;* for their opposition is asserted by the mind on account of some specific reason in the nature of the things considered. In the case of any collection of units that the number of them should be odd, and that it should be even, are contradictories consequentially related to the nature of odd and even integral numbers.

10. Secondly, we say, positively, that any two contraries become the contradictories of one another *when the circumstances of the case are such that the non-reality of either is the only condition wanting to complete an antecedent necessitating the reality of the other.* Evidently in any case the non-reality of either of two contraries is a condition of the reality of the other. Let this now be the only condition needed; thereupon the two contraries are contradictories. For whenever the non-reality of a first thing necessitates the reality of a second, the law of "the denied consequent" requires that the non-reality of the second must also involve the reality of the first.

In the case of a plane triangle there are three contraries; it may be either equilateral, or isosceles, or scalene. If now we limit the case to *triangles which have at least two sides equal*, only two contraries remain; and these are contradictories. For every triangle with at least two sides equal is either equilateral or isosceles. In general when a case of necessary consequence admits of two, and only two, alternative consequents, these become contradictories. If a house is certainly to be painted, and only two colors are obtainable, say brown and white, these are contradictories of each other.

11. Several things may be contradictory to one and the same thing. In a quadrilateral both the inequality of opposite sides and the inequality of opposite angles, are contradictories of its being a parallelogram. But things contradictory of one and the same thing cannot be contradictories of each other. For, being contradictories of one and the same thing, they must all be non-existent together if that thing exist; but things contradictory of each other cannot be non-existent together. Neither can the contradictories of one and the same thing be the contraries of one another: for they must all exist together if their common contradictory do not exist. Therefore we cannot have a series of mutual contradictories; as we can of mutual contraries. We must deal with contradictories in pairs.

The relation of contradictory to direct inference may be illustrated by the fact, that contradictory conceptions may always be found when two things are exact logical necessitants of each other. For either of such necessitants and the categorical contradictory of the other, are related to each other as consequential contradictories. To exemplify this, let smoke and fire involve the existence of each other, and the non-existence of either the non-existence of the other; then "smoke" and "no fire," or "no smoke" and "fire," are mutually contradictory. If we assert either we deny the other, and if we deny either we assert the other. The conception of contradictories, however, need not be based on that of necessitants; contradiction, like contrariety, can be perceived directly.

Moreover, as contrariety is specially noticeable between the species of the same genus, that is, between the specific forms of the same generic nature, so the most prominent mode of contradiction arises when a genus consists, or is made to consist, of only two species. In the case of integral numbers to be odd and to be even are natural contradictories; while to be odd and to be a multiple of four, are merely contraries. But should a collection of numbers contain only odd numbers and multiples of four, in that case, and with reference to the members of that arbitrary class, to be odd and to be a multiple of four would be contradictories.

12. When translative reasoning is based on the relations of contradiction, it is commonly expressed by the strong "*disjunctive syllogism.*" This consists of a major premise, setting forth the two contradictories in their double hypothetical opposition to each other; of a minor premise, in which one of the contradictory conceptions is actualistically asserted or denied; and of an actualistic conclusion. We say, "The line is either straight or bent," and then "it is straight, therefore it is not bent"; or "it is not straight, therefore it is bent"; or "it is bent, therefore it is not straight"; or "it is not bent, therefore it is straight." The disjunctive major premise is really a condensed statement of four hypothetical propositions; the minor actualistically asserts one of the four antecedents of those propositions; the conclusion is the consequent of that antecedent.

Evidently contradictory inference has two modes, the "*ponendo tollens*" and the "*tollendo ponens*": contrary inference has only one, the "*ponendo tollens.*" There is, however, a style of reasoning which might be called that of *contradictory contrariety*, in which, while dealing with contraries, we can regard them to some extent as contradictories; and can, therefore, use the "*tollendo ponens*" mode of argument, as well as the "*ponendo tollens.*" This arises *whenever the contraries in any given case are enumerated exhaustively;* and especially when a complete division is given of some existing genus. For instance, if we say, "The season was either spring, summer, autumn or winter," we not only can assert any one contrary and deny each of the rest, but, if we deny all the rest, we can assert that one; or, if we deny some and leave some undenied, we can assert these last disjunctively. For, if it is neither spring nor summer, it must be either autumn or winter.

In the syllogism, "The man is either a knave or a fool; he is not a knave, therefore he is a fool," the "*tollendo ponens*" holds good, although the major premise does not explicitly enumerate the contraries. The reason is that the conclusion is supported by a suppressed and understood contrary, the complete enumeration being, "The man is either a knave or a fool or *both*." We cannot, however, say, "The man is a knave, therefore he is not a fool," using the "*ponendo tollens*"; because the suppressed contrary does not support this conclusion. We see, therefore, that in every case of *tollendo ponens*, notwithstanding this apparent exception, all the alternatives are and must be considered. For in the foregoing instance the "*tollendo ponens*" is justified and the "*ponendo tollens*" condemned only after consideration of the suppressed alternative.

13. A complicated form of argument involving both direct and disjunctive hypothetical inference has been called the *dilemma*, or double assumption. Its major premise assumes two or more sequences as hypothetically true. Its minor premise is actualistic, and either asserts the antecedents of those sequences disjunctively, or denies their consequents disjunctively: in the former case the dilemma is "constructive"; in the latter, "destructive." The conclusion either asserts the consequents disjunctively, unless there be only one common consequent, in which case that is asserted; or it denies the antecedents disjunctively, unless there be only one common antecedent, in which case that is denied. The reasons for these operations are apparent from the nature of hypothetical and disjunctive inference. The constructive dilemma is either complex or simple, according as the sequences given in the major premise have different consequents or one common consequent; and, in like manner, the destructive dilemma is complex or simple, according as the given sequences have different antecedents or one common antecedent.

The following is a complex constructive dilemma:

"If a statesman who has discovered his policy to be mistaken, alters his course, he is chargeable with inconsistency; and if he do not alter it, he is guilty of deceit.
But he either does, or does not, alter it;
Therefore he must be either chargeable with inconsistency or guilty of deceit."

The following is a simple constructive dilemma:

"If a study furnish information, it should be pursued; and if it develop the mind, it should be pursued.
But this study either furnishes information or develops the mind;
Therefore it should be pursued."

The following is a complex destructive dilemma:

"If the man were wise, he would not speak irreverently of Scripture in jest; if he were good, he would not do so in earnest.
But he does it either in jest or in earnest;
Therefore he is either not wise or not good."

The following is a simple destructive dilemma:

"If the man were wise, he would not speak irreverently of Scripture in jest; neither would he do so in earnest.
But he does it either in jest or in earnest;
Therefore he is not wise."

In these last arguments it will be noticed that a negative consequent is denied by giving its contradictory; which is positive.

CHAPTER XVIII.

PROBABLE INFERENCE.

1. The tychologic principle. 2. Chances. 3. Their individuality. 4. Their arithmetical value. 5. Their addition and subtraction. 6. Their multiplication and division. 7. Compounded probability explained. 8. The compounding of a series. 9. The addition of compounded probabilities. 10. The application of the binomial formula to probabilities connected with recurrent trials. 11. Philosophical probability and improbability. 12. Probability may be either orthologic or homologic. 13. Analogical and inductive probability. 14. Moral certainty.

1. PROBABILITY attaches primarily to single inferences and to illative propositions as the expression of these inferences; after that, and in consequence of that, it may affect syllogisms, or those inferences which follow upon the composition of propositions. For if either premise of a syllogism be probable, the conclusion must be probable. If we can understand the nature of the single probable inference, no difficulty will be experienced respecting that of probable reasoning.

We find the radical law of all probable inference in the principle of "the ratio of the chances"; which principle may be named the *tychologic* principle.

2. The nature of "chances" is best explained by considering them, in the first instance, as conflictive consequents of possibility, and as making up a family, or company, of such consequents. The antecedent of an inference in possibility differs from the antecedent of an inference of necessity in that *the latter cannot have conflictive consequents, while the former may.* If a thing be necessary, nothing that cannot exist along with it can also be necessary. But two or more things may be possible at the same time, even while no two of them can be actual together. When we know simply that a book is a bound volume, we can say that it may be a quarto, or an

octavo, a duodecimo, or of some other style; but if it be any one of these, it cannot be any of the others. The fact that it is a bound volume is an antecedent of contingency, or possibility, with a number of conflictive consequents. Now, when a case admits of only a limited number of conflictive consequents, one of which must be realized, and when it presents no ground for believing that any one of them, rather than any other, has been, or will be, realized, we call the consequents "chances."

The relation between a chance and a consequent of necessity is such that the former changes into the latter whenever the antecedent of contingency is filled out, in any way, so as to make it an antecedent of necessity. Should we know not only that a volume is bound, but also that it is a copy of a book published only as a quarto, then we would say that the volume must be — not that it may be — a quarto. But, although chances are thus related to necessary consequents, they themselves are never necessary, or real, but only ideal, objects. When a chance is realized it ceases to be a chance; and its companions, also, cease to be chances in their failure to be realized.

3. Every chance is conceived of *as an individual, and not as a general*, consequent of contingency. Should a drawing take place from a box containing twenty black, thirty red, and fifty white marbles, there would be three general consequents of contingency, viz., that a black, that a red, and that a white marble, should be drawn. These general consequents would not be chances according to the logical use of language. A chance in the foregoing case would be the possible drawing of any one ball; and evidently there would be one hundred such possibilities. In determining the "ratio of the chances" we always conceive of these individual and equal possibilities. We cannot, however, always conceive of them as directly and as definitely as in the case just considered.

If a postman called at a certain house to deliver letters about four days out of every thirty, one might say at first that on any given day there are only two chances, viz., that he may, and that he may not, call. But properly these are two general events — or rather two events of a general character — each of which is supported by a number of chances. The case presents thirty chances — or individual possibilities of equal credibility. According to four of these the postman will call; according to twenty-six he will pass by. These chances may not be definitely conceived of in our judgment respecting the likelihood of a call; but they are the real rational basis for such a judgment.

The individuality of the chances means little more than that they are the units of measure among which the confidence of the mind is equally distributed. In every case of probability there is a necessity that one of the chances should be realized, and, as we have no reason to expect one more than another, we expect each equally, dividing among them the confidence of certainty.

4. In order to indicate the value of a single chance mathematically we must employ *a fraction whose numerator is unity and whose denominator is the whole number of chances.* In the case of the postman the value of each individual chance is one-thirtieth of certainty, while the two general events supported by the chances have the values $\frac{4}{30}$ and $\frac{26}{30}$, or $\frac{2}{15}$ and $\frac{13}{15}$. The application of mathematical methods to the determination of probabilities begins with this employment of fractions; and it leads to the addition, the subtraction, the multiplication and the division, of fractions representing degrees of probability.

5. When two or more events are specific forms of the same general event, so that, if either of them happen, that will be a happening of the general event, the probability of the general event is found by adding the probabilities of all the specific events included under it. Should a cubical die whose sides are numbered from 1 to 6 be thrown out of a box, the chance for any one number appearing uppermost is $\frac{1}{6}$, and the probability for the appearance of an odd number would be $\frac{3}{6}$, this being the sum of the chances for the three sides bearing the three odd numbers.

But were the die thrown twice, we could not say that the probability for an odd number appearing on both throws would be $\frac{3}{6} + \frac{3}{6}$, or unity; for "an odd number on both of two throws" would not be a general event possible in either of two specific forms, but a double event rendered possible by a doubled antecedent. We shall see that the probability of such an event is found by multiplication, not by addition.

Conversely, if an urn contain 30 white balls, and 70 colored red or otherwise, the probability for the drawing of a colored ball is $\frac{70}{100}$. And if fifty of these 70 chances favor other colors than red, the probability for a red ball must be $\frac{70}{100} - \frac{50}{100}$ or $\frac{20}{100}$ or $\frac{1}{5}$.

The foregoing examples illustrate the only cases in which the determination of probabilities calls for, and admits of, the addition and subtraction of fractions. These operations are applicable only when some general event and its specific forms are all conceived of as consequents of the very same antecedent of contingency.

6. The multiplication of probabilities — that is, of fractions indicating probability — is used *when one consequent of contingency, in other words, one probable event, is consequent upon another*. By means of this multiplication we determine the probability of the compound event which would be composed of both the probable events in case they should happen; which also is the probability of the conclusion, or of any other part, of this compound event, as part of it. If there be one chance in five that a certain criminal will be caught, and one in three that he will be convicted after being caught, it is plain that now, and until he may have been caught, the probability of his conviction will not be one-third of certainty — for that would involve the assumption that he certainly has been caught, or shall be — but only $\frac{1}{3}$ of $\frac{1}{5}$, or $\frac{1}{15}$; which also is the probability of the entire compound event of his being both caught and convicted. Therefore the probability of a compound event, or of any part of it as such, is the product of the probabilities of its component events.

Conversely, if we know both the probability of an event compounded of two probable events and the independent probability of one of its components, and if we divide the probability of the event by the probability of that component, we shall obtain the independent probability of the other component. If we know that the probability of a criminal being both captured and convicted is $\frac{1}{15}$, and the probability of his being caught is $\frac{1}{5}$, we can say that the probability of his being convicted, in case he is captured, will be $\frac{1}{3}$. Because $\frac{1}{15}$ divided by $\frac{1}{5}$ is equal to $\frac{1}{3}$.

7. In the foregoing illustration the events composing the compound event are related to each other as the condition of a result and the result conditioned. Such a relatedness, however, is not necessary in order to a compounding of probabilities. Any two events which are not of repugnant natures, and which, therefore, may both be realized, may be conceived of as one double event. The event of ace on one throw of a die, with the probability $\frac{1}{6}$, and that of head on one toss of a penny, with the probability $\frac{1}{2}$, may be compounded into the event "ace on one throw and head on one toss," with the probability $\frac{1}{12}$. It is evident, moreover, that, after two events have been compounded, a third may be compounded with the result; and that, in this way, any number of events may be made to compose an event whose probability is the product of the probabilities of its parts.

8. An interesting case of compounded probability occurs when the component events may be successively expected according to a regular series of fractions. After the shuffling of a complete pack of playing cards, the probability of a pictured card being uppermost is $\frac{12}{52}$, there being 52 cards in all, and 12 of these pictured. Then, should this event take place, and the pictured card be laid aside, the probability that the next card will be found to be pictured will be $\frac{11}{51}$. If this event occur and the second pictured card be laid aside, the probability for the appearance of a third will be $\frac{10}{50}$. And, if the subsequent drawings continue to yield pictured cards, the series will go on till only one such card remains in the collection; and will terminate with the fraction $\frac{1}{41}$, the probability for that card. Such being the separate probabilities for the successive appearances of pictured cards, that for a combined event can be easily determined. For example, the probability that the first three cards will be pictured must be the product

of the first three fractions of the series. That is $\frac{12}{52} \cdot \frac{11}{51} \cdot \frac{10}{50}$, or $\frac{1320}{132600}$, or a little less than one in a hundred.

But the probability that the first card will be pictured, the second plain, and the third pictured, will be the product $\frac{12}{52} \cdot \frac{40}{51} \cdot \frac{11}{50}$, or $\frac{5280}{132600}$, or about $\frac{1}{25}$. This would be the result also if the plain card were to be expected first, or last, of the three.

When the compounded probabilities are not a variable series, but equal to each other, as happens when precisely the same trial, or antecedent of probability, is repeated, the calculation is simpler. For instance, the probability that a pictured card will be uppermost three times in succession after three shufflings of the entire pack, would be found by raising the fraction $\frac{12}{52}$ to its third power. It would be $\frac{27}{2197}$, or more than one in one hundred.

9. A more complex class of problems than those hitherto noticed calls for *both the multiplication and addition of probabilities*. For addition is used whenever the probability of a general event is to be determined by uniting the probabilities of its specific forms.

Let the question concern the probability of obtaining "ace on two successive throws" of a die. This question is ambiguous; it may concern (1) the probability of "ace on both throws," or (2) that of "ace on one throw only," or (3) that of "ace on one throw at least," perhaps on both. In the first of these events, "ace on both throws," the second throw, by which the event may be completed, is not to be allowed unless the first throw is successful. We therefore compound the separate probabilities of these two throws, that is, we multiply $\frac{1}{6}$ by $\frac{1}{6}$ and find the probability required, $\frac{1}{36}$. This calls for no addition of fractions. But the event "ace on one throw only" may take place in either of two ways; for ace may appear on the first throw only, or on the second throw only; and the probability of it must be determined by adding together the probabilities of its specific forms. The probability of ace on the first throw and not on the second is $\frac{1}{6} \cdot \frac{5}{6}$, or $\frac{5}{36}$. That of ace not on the first throw but on the second is $\frac{5}{6} \cdot \frac{1}{6}$, or $\frac{5}{36}$. Adding these together we find the probability sought for, $\frac{10}{36}$ or $\frac{5}{18}$. Finally, the chances for "ace on one throw at least" (out of the two) comprise those of three possible events, viz. of "ace on first throw only," of "ace on second throw only," and of "ace on both throws." We have just seen that the united probability of the first two of these events is $\frac{5}{18}$. Add to this $\frac{1}{36}$, the probability of ace twice in succession, and we have $\frac{11}{36}$ as the probability of ace once at least in two throws.

10. An ingenious theorem respecting recurrent probabilities provides for the calculation of *the chances for an event happening any given number of times, in any given number of trials.*

Let an urn contain any number of balls, one third of them being red, and the rest of other colors. The probability that the first ball drawn out by a blindfolded person will be red is $\frac{1}{3}$; the probability that it will not be red is $\frac{2}{3}$. Moreover, if every ball drawn out be immediately replaced, these same probabilities will recur with every subsequent trial. We may now ask "What is the probability of a red ball appearing, say, 4 times, and failing to appear 6 times, in 10 consecutive trials? Or of its appearing 7 times, and failing to appear 3 times?" Such questions can be easily answered by the use of a mathematical formula.

Let us designate the event, the recurrence of which an exact number of times in a given number of trials is the subject of enquiry, by E, and its failure to occur by F, the probability of its occurrence on one trial by p, and the improbability of its occurrence on one trial by q. Then, on two trials, the possible combinations are the following double events: first, EE, with the compound probability $p \times p$, or p^2; FF, with the probability $q \times q$, or q^2; EF, with the probability $p \times q$, or pq; and FE, with the probability $q \times p$, or qp, or pq. Evidently no other combination than these is possible. Moreover, EF and FE, as they are alike constituted by one success and one failure, may be considered as varieties of that general compound event *in which we conceive of one event and one failure without respect to order of occurrence.* The probability of that event, therefore, will be $pq + pq$, or $2pq$, this being the sum of the probabilities of the specific events. If, now, we drop either the designation EF or FE and use the other (say EF)

for that general event which covers both EF and FE, we shall have only three possible events, EE, EF, and FF, with probabilities expressed respectively by the terms of the polynomial $p^2 + 2pq + q^2$. For example, $\frac{1}{6}$ being the probability for ace on one throw of a die, and $\frac{5}{6}$ the probability for the failure of that event, these values being substituted for p and q in the foregoing polynomial, the several terms give the probabilities for ace twice on two throws, EE; for ace once only on two throws, EF; and for failure of ace on both throws, FF. Thus — $(\frac{1}{6})^2 + 2(\frac{1}{6} \times \frac{5}{6}) + (\frac{5}{6})^2$, or $\frac{1}{36} + \frac{10}{36} + \frac{25}{36}$.

It should be noticed that the sum of these fractions is unity, or one; as it ought to be. For, since one or other of the three events must happen, they divide between them all the chances in the case.

Should we now make three trials, instead of two, the possible compound events, with their probabilities, will be as follows:

EEE with the probability $ppp = p^3$
EEF " " " $ppq = p^2 q$ ⎫
EFE " " " $pqp = p^2 q$ ⎬
FEE " " " $qpp = p^2 q$ ⎭
EFF " " " $pqq = pq^2$ ⎫
FEF " " " $qpq = pq^2$ ⎬
FFE " " " $qqp = pq^2$ ⎭
FFF " " " $qqq = q^3$.

Again disregarding the order in which the component events may occur, these eight compound events may be conceived of as four, namely, EEE, EEF, EFF, and FFF; and, evidently, the probabilities of these four events are expressed by the terms of the polynomial, $p^3 + 3p^2 q + 3pq^2 + q^3$. This is the cube, just as the polynomial first obtained was the square, of the binomial $p + q$. In like manner the development of this binomial to its fourth power, will give the different probabilities that an event with the separate probability, p, will occur, on four trials, every time, or only thrice, or only twice, or only once, or not at all; — the antecedent of probability being exactly repeated in every trial. And, in general, the development of $p + q$ to the nth power will give all the probabilities of the occurrence of an event any number of times on

n trials; p being the probability, and q the improbability, of the event on one trial. For instance, in order to determine the chances for "ace three times only in ten throws," we must raise $p+q$ to the 10th power, and then find the numerical value of the term whose literal part is $p^3 q^7$, after substituting $\frac{1}{6}$ for p and $\frac{5}{6}$ for q.

The foregoing theorem belongs to a department of mathematics in which men of genius have discussed many interesting problems, and which may be taken as a proof that problematic inference is no less rational in its methods, and no less connected with the necessary nature of thought and of things, than apodeictic inference is.

In the above discussions the "multiplication" of probabilities must be taken to mean their multiplication one by another — the compounding of them. The multiplication of a probability by a whole number is an operation altogether different from this: it is only a short way of adding the equal probabilities of two or more possible specific results, connected with one and the same antecedent, and whose united probability is that of a more general event.

11. The calculation of chances brings into prominence *a wide philosophical use of the words "probable" and "improbable"*; which it may be well to define. Ordinarily the probable is that which has the majority of the chances in its favor; and the improbable is that which only a minority of the chances support. According to common speech the same event cannot be both probable and improbable under the same circumstances. But, philosophically, that is probable which has any chances at all in its favor, whether they be few or many; and that is improbable which has any chances at all against it. According to these technical meanings the most improbable event has some degree of probability, and the most probable some degree of improbability. In ordinary language we say that "ace on the first throw of a die" is not probable, but highly improbable; mathematically and philosophically, it has the probability of one-sixth.

Probable as well as demonstrative inference may take place either orthologically — that is, directly from an antecedent, and without reference to any previous case of similar consequence — or homologically. In this latter case, instead of directly estimating chances, we give the consequent of a repeated antecedent the same probability as in a previous judgment. The instances of probable inference discussed above are all orthologic; but any of them may be the basis for an homologic inference.

12. The relation between orthologic and homologic inference is precisely the same in problematic as in apodeictic sequence; and does not call for special discussion. Some remarks, however, are in place here concerning our probable judgments respecting natural laws and events. These all pass under the generic name of probable induction, but are often divided into two classes, one of which is especially entitled to this name, and the other of which is sometimes known as reasoning from, or according to, the analogy of Nature. Probable induction, in the specific sense, is essentially a form of principiation; and takes place when a consequent follows a causational antecedent, not invariably, but only sometimes, or for the most part. This happens when some power adequate to produce a result is occasionally counteracted; or when some tendency which needs advantageous circumstances to render it effectual, is only sometimes attended by such circumstances. Ordinarily a wound produces pain; mental excitement or bodily stupor or rapidity of infliction may prevent this. Therefore we only say that a wound is likely to produce pain.

Three steps may be distinguished in forming this inference. We say, first, "Most wounds have caused pain"; then, by principiation, "Most wounds produce, or will produce, pain"; and then, by the tychologic principle, that is, according to the ratio of the chances, "A wound is likely to produce pain." Because any wound, taken at random, may be one of those which are painful. But, in the foregoing process, the tychological judgment may precede the principiation without any change in the result. Thus, having seen that most wounds have produced pain, we may say, first, that any one of these observed wounds was probably painful, and then, inductively, that any wound whatever is likely to be painful. Ordinary judgments of probability are formed in one or other of these ways.

13. The inference from the "analogy of Nature" differs from the foregoing in that it is not supported by the sameness, or exact similarity, but only by an imperfect similarity, of the new antecedent to that already known to have a certain consequent. It is founded on the principle that whatever in the natural Universe resembles a known cause, or reason, is probably or possibly a true and sufficient reason. Though this inference may terminate in principiation it is mainly paradigmatic; it is essentially a reasoning from one parallel case to another; but it is founded on a parallelism which has been only imperfectly established; and which, therefore, is only probable, or not unlikely. Both the inductive and the analogic inference assume confidently that Nature has an intellectual unity, and that her methods are fixed and uniform: but in the former case a law of natural action has been discovered, while in the latter there are only grounds for conjecture. The probability of probable induction arises because, though the antecedent is defined and known, the consequent does not follow always, but with exceptions; that of analogical inference arises because the existence of a sufficient antecedent is only a matter of probability or of possibility. We cannot be sure that other planets or stars are inhabited because our world contains the race of Adam; for the case presents only a probable or conjectural antecedent.

14. When the proportion of the chances in favor of any event, or course of events, is so great that we feel authorized utterly to disregard the chances against it, the event is said to be morally certain. Thus the alternation of day and night, and of the seasons, and the continued earthly existence of the human family, during the coming year, are things of which we have no doubt. For the practical purposes of logic this certainty differs nothing from the conviction of clear memory, or of immediate cognition, or of demonstrative evidence. But it is important to remark that *the highest probability can never*

really reach absolute certainty. No matter how extreme the likelihood of a thing may be — no matter how small the proportion of the chances against it to the chances for it may be — still, so long as a thing is probable, there is a possibility of the opposite. Were there a thousand millions of chances for an event, and only one against it, that one chance would involve the perfect possibility of its non-occurrence.

CHAPTER XIX.

THE OPPOSITION OF PROPOSITIONS.

1. Pertains to illative propositions having the same subject and predicate. 2. Dogmatic assertions may be opposed in quantity and in quality. 3. Their subalternation and superalternation. 4. Their contrariety. 5. Their contradiction. 6. Exclusively partitive assertions. 7. Subcontrariety. 8. Summary of the laws of opposition. 9. Modal opposition essentially corresponds with dogmatic. 10. Modal contrariety. 11. Modal subalternation. 12. Modal contradiction. 13. This last kind of opposition involves a specific kind of contingency in the sub-contraries. 14. Modes of contingency or possibility: (*a*) the embedded, (*b*) the unstable, or unguarded, (*c*) the stable, or guarded, (*d*) the half-stable, or half-guarded. 15. This last, as positive and negative, becomes encouraging and discouraging contingency; and furnishes the contradictories of necessity and impossibility. 16. Sub-contrariety. 17. Probability participates in the oppositions of contingency.

1. ILLATIVE propositions are the most important in logic. Among illative propositions those which are general, and which, therefore, present laws, or rules, of inference, are especially important. Logic is so much concerned with these that it might even be called the science of "Canonics"; as it was anciently by the Epicureans.

These general illative propositions, when expressed categorically, are of two classes; first, the pure, or dogmatic, in which we make either universal or particular assertions respecting the members of a logical class; and secondly, the modal, or conditionative, in which we make either apodeictic or problematic assertions respecting a general subject.

When two propositions of either of these classes have the same subject and predicate terms, but differ otherwise, they are said to be opposed — immediately opposed, to one another. Sometimes, with a restricted use of language, only propositions which are contrary to, or contradictory of, each other, are spoken of as mutually opposed. But logicians also characterize any propositions as being opposed to each other when they have the same subject and predicate terms, but differ otherwise; whether they conflict with each other or not.

The various modes in which propositions may be opposed are worthy of study, because, in every opposition, the truth or the falsity of at least one of the opposed propositions can be inferred from the truth or from the falsity of the other. Let us enquire first concerning the oppositions of pure, and then concerning those of modal, categoricals.

2. Dogmatic categoricals may be opposed *in quantity, or in quality, or in both.* They are opposed in quantity when one is universal and the other particular, both being either affirmative or negative; in quality when one is affirmative and the other negative, both being either universal or particular; in both quantity and quality when one is universal and affirmative and the other particular and negative, or when one is universal and negative and the other particular and affirmative. The propositions "all men are wise" and "some men are wise," as also the propositions "no men are wise" and "some men are not wise," are opposed in quantity. The propositions "all men are wise" and "no men are wise," as also the propositions "some men are wise" and "some men are not wise," are opposed in quality. The propositions "all men are wise" and "some men are not wise," as also the propositions "no men are wise" and "some men are wise," are opposed in both quantity and quality.

For the sake of brevity logicians indicate these different forms of opposed propositions by the vowels $A \, E \, I$ and O: A stands for the universal affirmative, E for the universal negative, I for the particular affirmative, and O for the particular negative.

The oppositions of these propositions have also been symbolized by the sides and the diagonals of a square, the corners of which have been marked severally by the four vowels. In short, the eye is made to help the mind, by means of the following figure, which is called "the logical square":—

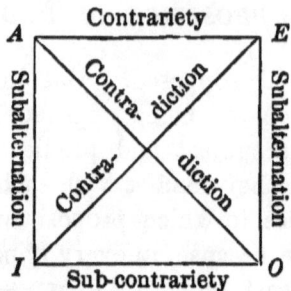

Reference to this diagram will be found to assist the apprehension and the memory.

3. The opposition of subalternation (which also, from a less important point of view, may be termed superalternation) exists between *A* and *I* and between *E* and *O*. In this opposition, according to the common view, the subalternate follows if the superalternate be allowed. If "all men are wise," then "some men are wise"; if "no men are wise," then "some men are not wise." These are immediate orthologic inferences; the law governing them is "Aristotle's dictum," that "whatever is true of a class universally, is true of any number of its members." But this axiom does not authorize the converse inference; we cannot infer the superalternate from the subalternate, as such. We can, however, on the principle of the "denied consequent," infer the falsity of the superalternate from the falsity of the subalternate.

Here, however, we must add that the relation of subalternation may be defined in a way which implies that the subalternate cannot be completely and exactly inferred from the superalternate. This definition is an improvement on the common view of subalternation, and will be explained later, in the present chapter.

4. The opposition of contrariety, or confliction, exists between *A* and *E*. If *A* be true, *E* is false; and if *E* be true, *A* is false. If "all men are wise," it cannot be that "no men are wise"; and if "no men are wise," it cannot be that "all men are wise." This opposition is a consequence of that immediate contrariety which exists between each superalternate and the denial of its subalternate. *E* immediately involves

the denial of *I*, which denial is immediately contrary to *A*; therefore *E* involves the denial of *A*. If "no men are wise," it is not true that "some men are wise"; and, if it is not true that "some men are wise," it is not true that "all men are wise." In like manner, *A* is immediately followed by the denial of *O*, and the denial of *O* by the denial of *E*. *A* and *E*, therefore, are the contraries of one another.

But *E* is not the only contrary of *A*, nor *A* of *E*. Each has another — a co-ordinate contrary. For *A* must be false if only *O* be true, while *E* is false; and *E* must be false if only *I* be true, while *A* is false. In either of these cases *A* and *E* are false together. *A* and *E*, therefore, are only contraries, not contradictories; this latter name being reserved for propositions between which the opposition is more thorough-going.

5. This "contradictory opposition" takes place between *A* and *O*, and also between *E* and *I*. We can say, "If *A* be true, *O* is false; and if *O* be true, *A* is false: also, if *A* be false, *O* is true; and if *O* be false, *A* is true." *E* and *I* contradict each other in the same way.

To explain this thorough-going contradiction we must particularly note that the designation "some," which belongs to *I* and *O*, indicates *an indefinite number which may include the whole class*. It does not mean "some only," or "some, not all," but "some at least," "some, perhaps all." Only that unrestricted "some" which may prove to be "all," can form the contradictory of an universal statement. For, in case there be a class of beings called "men," one of three contrary alternatives must follow concerning any characteristic of them — say, wisdom. Either "all men are wise," or "some, not all, are wise" (in other words, "some are wise and some are not wise"), or "no men are wise." Now the indefinite statement designated by *O*, that is, "some men — some at least, perhaps all — are not wise" is a general alternative including both the *second* and *third* of those just given as the contraries of the universal affirmative. It may, therefore, be regarded as the only alternative of *A*. But when one or other of two conflicting alternatives must exist, these are the contradictories of each other. Hence *A* and *O* are contradictories. In the same way E and I are contradictories. For I is a general alternative including the *first* and *second* of the three mentioned above.

6. Since the "some" of I and O does not preclude universality, we cannot argue from I that A is not true, nor from O that E is not true. Sometimes, however, as we have seen, "some" has a strictly partial, or partitive, meaning. In that case the partitive affirmative and the universal affirmative are contraries; and so are the partitive negative and the universal negative. For if "only some men are wise" it is not true that "all men are wise," and if "only some men are not wise" it is not true that "no men are wise." These partitive, or exclusively partial, propositions, are worthy of notice, yet need not be given any formal place in logic. For every such statement is compounded from two particular predications of the ordinary kind, made at once and in modification of each other. Should we, in saying, "Some animals are oviparous," so emphasize "some" as to signify "some only," this would be equivalent to saying, "Some animals are oviparous and some are not oviparous"; and the effect of this double statement would be the same as that of I and O operating together.

7. Finally, the opposition of sub-contrariety exists between I and O. These are styled subcontraries, partly because they are subalternates to the contraries A and E, but also because they have a peculiar contrariety of their own. For, while we infer the falsity of either contrary from the truth of the other, but not the converse of this, so we may infer the truth of either subcontrary from the falsity of the other, but not the converse of this. Contraries cannot both be true, though they may be both false; subcontraries cannot both be false, though they may be both true. If it be false that "some men are wise," it is true that "no men are wise"; and this justifies the subalternate "some men are not wise." If it be false that "some — or any — men are not wise," then "all men are wise"; and this justifies the subalternate "some men are wise." Inasmuch, however, as we seldom use a particular conclusion when the antecedent warrants an universal assertion, the inference of the truth of one subcontrary from the falsity of the other occurs but seldom.

8. Comparing the different modes in which pure categorical propositions may be opposed to one another, we find the most important to be that of contradiction. This yields the following sequences:

> A true, then O false; A false, then O true.
> E true, then I false; E false, then I true.
> I true, then E false; I false, then E true.
> O true, then A false; O false, then A true.

Next in importance is contrary opposition. This yields:

> A true, then E false; E true, then A false.

But there is no sequence, in contrariety, from A false, or E false.
Next subordination, or superalternation, yields:

> A true, then I true; E true, then O true.

But it gives no necessary conclusion from I true, or O true.
Finally, subcontrariety gives:

> I false, then O true; O false, then I true.

But it yields no necessary sequence from the assertion of I or of O.

Examining critically all the foregoing modes of sequence it appears that the truth of an universal assertion (whether A or E) involves either the truth or the falsity of every one of the three propositions which may be opposed to it. In like manner the falsity of a particular assertion (whether I or O) determines the truth or the falsity of every one of the three propositions which may be opposed to it. Therefore to assert an universal and to deny a particular are the strongest modes of predication. The weakest modes are those asserting the falsity of an universal or the truth of a particular; because each of these determines only one out of the three opposed judgments.

9. If now we turn to the mutual opposition of modal categoricals, we shall find in it an essential correspondence to that of pure categoricals, and at the same time, peculiarities arising from the fact that it pertains to more abstract and searching modes of thought. The universal dogmatic proposition ex-

presses necessity, either positive or negative; and the particular dogmatic proposition expresses a contingency, either positive or negative. In conformity with this we find, in modal propositions, contrariety between the necessary and the impossible; subalternation between the necessary and the possible to be, and between the impossible and the possible not to be; contradiction between the necessary and the possible not to be, and between the impossible and the possible to be; and subcontrariety between the possible to be and the possible not to be.

In this statement the terms "necessary" and "impossible" signify the necessary to be and the impossible to be, this latter being only the necessary not to be, viewed in a peculiar way; just as the necessary to be is the impossible not to be.

Let us take the propositions, "The tea-plant must — or certainly will — grow," "The tea-plant cannot grow," "The tea-plant may grow," and "The tea-plant may not grow"; noticing that in this last the negative particle does not qualify "may," but attaches the idea of non-existence to "grow." These assertions cannot all be true under the very same antecedent, or set of conditions; yet each must be true provided its own proper antecedent exist. For, (1) in cases yielding all the needful conditions of soil, climate, cultivation, and so forth, the tea-plant must grow; (2) in cases where any of the needful conditions is wanting, it cannot grow; (3) in cases where some conditions are known, but others are not known, to exist, we say that it may grow; and (4) in cases where some of the conditions are not known to exist, though some are known to exist, we say that it may not grow. Let us designate these four styles of assertion by the accented letters A' E' I' and O'.

10. A' and E' are contraries. They cannot be true together, and they may be false together. With the same antecedent it cannot be both necessary and impossible that the tea-plant should grow. Moreover, the facts obtainable in the case may justify neither of these judgments, but only some form of contingency. For in all these judgments the antecedent is *the tea-plant considered as in some set of circumstances or other;*

and in a contingent judgment the antecedent would be the tea-plant in the cases marked (3) and (4); which are, in truth, but two aspects of the same case. With this antecedent both A' and E' would be false, that is, unwarranted, judgments; as they would be after any antecedent which merely justified contingency. If we knew simply that a man was sick, it would be a false judgment to say either that he must die or that he cannot die. The antecedent only warrants "he may die" and "he may not die." For some sick men die and some do not. If, indeed, the man is found to have consumption, we may say, "He must die"; or, if he has but a headache, we may say, "He certainly will not die"; but in these cases new antecedents are used. The simple antecedent of "sickness" does not make death either necessary or impossible.

11. A' and I', as also E' and O', are respectively superalternate and subalternate. For there is a sense in which whatever is necessary is possible to be, and whatever is impossible is possible not to be. Whatever must be so, may be so; whatever cannot be so, may be not so. In the former case existence, and in the latter non-existence, consists with given surroundings. Yet this inference of contingency from necessity, or from impossibility, is only partial. For possibility and contingency, in the ordinary and proper sense of these terms, involve the possibility of the opposite. They are perceived when some of the conditions of a thing are known to exist, and some are not known to exist. Only the first of these things follows from a known necessity; only the latter from a known impossibility. Contingency, therefore, is said to follow from necessity, *only because necessity implies the positive part of the foundation of contingency;* and it is argued from impossibility *only in that impossibility implies the negative part of the foundation of it.* These, however, are the elements which give importance to positive and negative contingency respectively.

The inference of the subalternate particular from the universal, in pure categoricals, *has this same partial and one-sided character.* We cannot fully infer "some, perhaps all, are," from "all are," nor "some, perhaps all, are not" from "none are." The "some" part follows, but the "perhaps" part does not. Indeed, as the "perhaps" implies the possibility, or contingency, of "not all," it really conflicts with the universal judgment.

The inference of subalternation shows only that the particular — or the contingent — proposition has been *in the right direction*, not that it has expressed the full and exact truth: and *as subordinate to the universal and the necessary*, the particular and the contingent cannot be taken to mean that the superalternate proposition may not be true. They set forth only that peculiar particularity and possibility — or partitiveness and contingency — which are not apposed to, but involved in, the universal and the necessary.

12. Again, the modals A' and O' are the contradictories of each other, as are also E' and I'. For if it is true that *a thing must be so*, it is false that *it may not be so;* and if it is true that *a thing may not be so*, it is false that *it must be so:* also, if it is false that *a thing must be so*, it is true that *a thing may not be so;* and if it is false that *a thing may not be so*, it is true that *it must be so.* Likewise, if it is true that *a thing cannot be so*, it is false that *it may be so;* and if it is true that *a thing may be so*, it is false that *it cannot be so;* also, if it is false that *a thing cannot be so*, it is true that *it may be so;* and if it is false that *it may be so*, it is true that *it cannot be so.*

13. In these inferences, however, two peculiarities are to be observed in the significations of "may not" and "may" — two modifications of meaning which are not always, or necessarily, attached to these words.

First, when the falsity of one contradictory follows from the truth of the other, "may" denotes an absolute possibility to be — a possibility which cannot be displaced by impossibility; so that, with the given antecedent, the thing is certainly not impossible: while "may not" signifies an absolute possibility not to be; so that — the antecedent remaining without addition or alteration — the thing is certainly not necessary.

Secondly, when the truth of one contradictory is inferred from the falsity of the other, the word "may" is not used simply, but means "may, perhaps must" (equivalent to "may or must"), while "may not" means "may not, perhaps cannot," (equivalent to "may not or cannot").

Without these significations of "may" and "may not" the inferences of contradiction would not take place. The force of these words as thus modified may be inadequately expressed by saying that "may" means "certainly may, possibly must," while "may not" means "certainly may not, possibly cannot." These meanings — it will be noticed — correspond exactly to the "some, perhaps all," and the "some, perhaps none" of the dogmatic subcontraries. But they belong to a wider and more searching range of thought.

---

14. In some of the foregoing statements the term contingency has been used interchangeably with possibility, the reason being that contingency is based on possibility, and by reason of its radical nature, shares in the oppositional relations of possibility. As regards subalternation and contradiction the possible and the contingent are one; and, as we have seen, each of these modes of opposition may be said to involve a specific form of possibility of its own.

A clearer understanding of these teachings can be had if we consider four styles of possibility, in all of which the essential conception of the consistency of the existence — or of the non-existence — of a thing with given circumstances, is modified by some addition.

First of all, there is *that possibility which is perceived as accompanying necessity, positive or negative* — in other words, as accompanying necessity and impossibility. That which necessarily exists, is recognized as possible to be; and that which necessarily does not exist, is recognized as possible not to be. That which must be, may be; and that which cannot be, may not be. The positive form of this possibility may be said to be embedded, or infixed, in necessity; and the negative form to be embedded in impossibility. The positive form, therefore, cannot co-exist with impossibility, nor the negative with necessity. Yet a denial of the positive form does not warrant the assertion of impossibility, nor a denial of the negative form the assertion of necessity. Each form exists only in its own mode of necessity, positive or negative; and neither exists in case a given antecedent supports neither necessity nor impossibility. If death were considered possible only because death is necessary, a denial of this embedded possibility would not involve the impossibility of death. The case might be one of a possibility lying between necessity and impossibility, and not embedded in either.

Therefore this embedded possibility is not that according to which the contingent negative contradicts the necessary, and the contingent positive the impossible, but merely that according to which the contingent is inferred from the necessary and the impossible. For an antecedent of necessity, positive or negative, is taken as proof that the existence — or the non-existence — of a thing is consistent with a given set of circumstances. We allow that the possibility in such a case is only partial and improper. Ordinarily when we say that a thing is possible we do not mean that it is possible only to be, but that, so far as our knowledge extends, it is possible *either to be or not to be*. So also contingency commonly excludes the assertion of necessity. But if subalternation be taken as a mode of sequence in which the particular follows the universal and the contingent the necessary, it must be explained in the foregoing way; and with some such use of terms.

This mode of opposition, however, may be interpreted to mean that the *ordinary* particular or contingent assertion — the "some, perhaps all" or the "may, perhaps must" — is to be accepted as partially correct and as made in the right direction, provided the apodeictic superalternate be true. We really prefer this explanation, though it will not permit us any longer to infer the subalternate from the superalternate, but only that the subalternate has a certain logical value.

A second, and an entirely proper, form of possibility is recognized, *when the very same antecedent which supports a contingency may, or may not, support a necessity, yet is not perceived either to do so or not to do so*. If one knew only that a lion is a quadruped and that a quadruped may and may not be a carnivore, he could say, "A lion may be a carnivore," and, "A lion may not be a carnivore." In so doing he would use the antecedent "lion" correctly, but without knowing the full force of it; because, absolutely speaking, the lion is necessarily a carnivore. Or should one use an antecedent capable only of supporting contingency, while he yet knew not of this limitation, this would be another species of the kind of inference now mentioned. If one knew only that a merchant is a man, and that a man may and may not be wise, he could say, "A merchant may be wise," and, "A merchant may not be wise." But he could not say whether further knowledge might not warrant "a merchant must be wise" or "a merchant cannot be wise." The contingency thus described may be styled "*unstable*," because further knowledge of the antecedent may lead one to displace the contingent by an apodeictic judgment. It may also be called "*unguarded*," because it is not protected, as another form of contingency is, against being displaced by necessity or impossibility. Evidently no unstable contingency can contradict an apodeictic statement; since the latter may prove to be supported by the very same antecedent which is employed to support the contingency.

A third form of possibility or contingency is *the stable, or guarded*. This is recognized when the antecedent is perceived to be of a nature to support contingency only; so that no further knowledge of the antecedent can justify a judgment in necessity or in impossibility. In this case we say that a thing is neither necessary nor impossible, but possible to be and possible not to be.

Every judgment in contingency may, on further information, be displaced by an apodeictic judgment. This may happen to an unstable contingency while the antecedent remains the very same; but it cannot happen in stable contingency so long as the antecedent be not essentially modified, or replaced by a new antecedent. That ace may appear and may not appear on the throw of a die, and that frost may and may not occur on New Year's in latitude 40°, are stable, or guarded, possibilities; they cannot, with the antecedents given, become certainties.

Stable contingency is also perceived when the antecedent has been seen to be *sometimes accompanied, and sometimes not accom-*

panied, by the consequent. Knowing that man sometimes is wise and sometimes not wise, we assert, as a stable contingency, "man may be, and may not be, wise." The observed instances preclude us from saying that man is necessarily wise, or that he is necessarily not wise. It is especially when determined in this way that stable contingency may be called guarded.

This contingency — that is, the assertion of it — denies both necessity and impossibility; because neither necessity nor impossibility can result from one of its antecedents. By it the possible to be is opposed to impossibility, and the possible not to be to necessity; and so the positive side of it contradicts impossibility, and the negative, necessity; but the force of the contradiction comes from that stability which affects both sides alike and together. In the same manner necessity and impossibility contradict this contingency; that is, are the contraries of it.

But a denial of stable contingency does not compel the assertion of necessity or of impossibility; it only involves that *one or other* of these is true. The denial of necessity, moreover, involves only that the contingency or impossibility is true — not that the contingency is true; and the denial of impossibility involves only that the contingency or the necessity is true, not that the contingency is true. Thus the stable contingency, "a quadruped may, and may not, be a carnivore" conflicts with both positive and negative necessity; as each of them does with it. But were it false, this would not justify us in saying either that a quadruped must be, or that a quadruped cannot be, a carnivore; we could only say that one or other of these is true; nor would the falsity of one only of these apodeictic statements justify the assertion of the contingency. Therefore *thorough-going contradictory opposition to necessity or to impossibility cannot be obtained from stable contingency.*

There is, however, a fourth style of contingency which does yield this opposition. It may be called *half-stable*, or *half-guarded*, contingency. It is perceived when an antecedent of contingency has been seen sometimes to be actually followed by a consequent, and has never been seen without the consequent; or when an antecedent has been seen sometimes to occur without the consequent, while it has never been seen to be followed by it. In the former case the contingency is guarded against impossibility, but may be displaced by necessity; in the latter it is guarded against necessity, while it may be displaced by impossibility.

15. A half-stable contingency guarded against impossibility may be said to lean, or tend, towards necessity. It is expressed exactly, though indirectly, by a particular affirmative asserted alone. Thus, "some men are wise" means "*man may be wise, perhaps must be, and certainly is not incapable of wisdom.*" A half-stable contingency guarded against necessity leans, or tends, towards impossibility; and is expressed similarly by the particular negative. Thus, "some men are not wise," as an isolated statement, means "*man may not be wise, perhaps cannot be, and certainly is not necessarily wise.*"

No terms have been used to designate these two modes of half-stable possibility. Let us call that which leans towards necessity *encouraging* possibility, or contingency; and that which leans towards impossibility *discouraging* contingency. These somewhat arbitrary terms are the best which suggest themselves.

We are prepared, now, to state what forms of contingent assertion are the thorough-going contradictories of necessity and of impossibility. *Necessity is thoroughly antagonized by discouraging contingency, and impossibility by encouraging contingency.* If we assume simply that "some men are wise," we can assert the encouraging contingency "man may be wise." Then we can say, if it is true that "man may be wise," it is false that "man cannot be wise"; if it is false that "man may be wise," it is true that "man cannot be wise"; if it is true that "man cannot be wise," it is false that "man may be wise"; and if it is false that "man cannot be wise," it is true that "man may be wise." In short, encouraging contingency and impossibility thoroughly contradict each other; and so do discouraging contingency and necessity.

The main object of the foregoing discussion has been to bring out the inner nature and meaning of *those particular prop-*

ositions which are the thorough-going contradictories of universals. Evidently they are at heart a peculiar kind of contingent modals. Hence, too, it should be noticed that the symbols I' and O', as used to indicate the contingent equivalents of particular propositions, relate only to half-stable contingency, and not to contingency in general. The prominence thus given to half-stable contingency is not unreasonable: dialectically this is the strongest mode of contingency; and it is of peculiar value when we are seeking the actual, and not merely the possible or the probable.

16. Finally, the relation of subcontrariety exists between the propositions just described, that is, between I' and O', as half-stable contingencies. For if either of these be false the other is true. Yet, critically speaking, this sequence does not take place exactly. The exact sequence is from the falsity of I' or of O' to the truth of the opposite *embedded* contingency. A corresponding inaccuracy appears when we say that the falsity of a particular proposition involves the truth of the opposed particular. For the falsity of "some, perhaps all, are," does not involve the doubtful conclusion that "some, *perhaps* all, are not," but the absolute conclusion that "some, *as a part of all*, are not"; and the corresponding inference from the falsity of "some are not," is that "some, *as a part of all*, are." The falsity of one subcontrary shows that the other has been made in the right direction, though it has fallen short of the truth; this is all that the logical rule can be taken to mean, whether the propositions be pure or modal. For the subcontraries are really, not embedded, but half-guarded, contingencies.

These modal subcontraries agree also with the pure subcontraries in that the truth of either does not involve either the truth or the falsity of the other. That "some (perhaps all) men are wise" does not involve that "some (perhaps all) men are not wise," because the first of these statements implies that we may find all men to be wise, in which case "some men are not wise" would be entirely false. But the two particular propositions will be true together, *so far, at least, as regards their* "*some are*" *and* "*some are not*," in case we do not find all men to be wise. In like manner, "man may be, perhaps must be, wise" does not involve "man may not, perhaps cannot, be wise," because, if man should prove to be necessarily wise, this would show that the discouraging contingency had been falsely asserted. Indeed, in the strictest sense, encouraging and discouraging contingency conflict with each other. Yet, should we find that, though man may be wise, he is not necessarily wise, then both the positive and the negative contingency would be true so far as the "may" and the "may not" are concerned.

17. Probability has not been mentioned in the above discussion. The oppositional relations of this mode of sequence are essentially those of possibility; and belong to probability as being based on possibility. Probability presupposes some form of contingency proper; and may be divided into unstable, stable, and half-stable, according to the style of contingency on which it is based. In unstable probability the ratio of the chances is determined provisionally and temporarily; because the very same antecedent which yields probability may be found to yield certainty. In half-stable probability the ratio of the chances is guessed at roughly; because our knowledge extends only to instances favoring one side. Permanent, duplicating, or recurrent probability, which is the leading form of this mode of assertion, is "stable"; and as such, while not justifying either the subalternation or the thorough-going contradiction of judgments, conflicts with both necessity and impossibility.

The importance of modal opposition relates to the contradictions between possibility (including contingency) and necessity; it is not connected with the specific nature of probability.

CHAPTER XX.

THE CONVERSION OF PREDICATIONS.

1. The conversion of dogmatic, or "pure," propositions. 2. Requires a substantialized and quantified predicate. Proceeds on the principle of *Identity*. 3. The ordinary conversion of affirmatives. 4. And of the universal negative. "Not," as the exclusive particle. 5. Particular negatives must be converted by "contraposition," or "infinitation." 6. Conversion "per accidens," or "by limitation." Conversion "per differentiam," or by "the retained-necessitant." "Simple" conversion. 7. The quantification of modals. 8. The universal-necessary and the particular-contingent. 9. The universal-contingent. 10. The particular-necessary. 11. Quantity is non-essential to modal thought. The ordinary converse of a necessity is a simple contingency; but sometimes we convert with the retained-necessitant. 12. The ordinary converse of an impossibility is an impossibility. 13. The conversion of contingency. Always follows "the asserted-consequent." 14. Contingent and particular conversion compared.

1. CONVERSION, or, more expressly, conversional sequence, takes place whenever from a given proposition another is inferred in which the same terms appear but with an exchange of places. Like oppositional sequence it is not dependent on any reference to a previously perceived similar sequence; and is, therefore, orthologic. The antecedent proposition is called the convertend; the consequent proposition, the converse. The subject of the convertend furnishes the predicate of the converse, and the predicate of the convertend the subject of the converse. For example, from "all men are mortal" we infer that "some mortals are men."

Propositions purely factual, or historical, may be converted. Because "Mr. Harrison is president elect," we can say "the president elect is Mr. Harrison." From "some rogues are on that jury" it follows that "some on that jury are rogues." Such inferences not only follow the law of Identity (Chap. XV.), but are entirely explained by means of it: no study is required to understand them. The conversion now to be discussed pertains to those illative propositions which may be used as principles in reasoning, and especially to categorical predication as expressing general hypothetical sequence. Let us consider, first, the conversion of pure, or dogmatic, categoricals; and, after that, the conversion of modal predications.

Before commencing this discussion it should be observed that not all propositions are convertible; only those which have been distinguished as inherential statements, or predications proper. Presentential propositions cannot be converted, because they never set forth a sequence, nor any relation between things, but merely the existence or the non-existence of the subject.

2. The conversion of a dogmatic predication takes place *only after the predicate of the convertend has been both substantialized and quantified.* Substantialization is effected when the predicate is changed from the ascriptional form of thought, such as adjectives and verbs express, to the substantial form, which is expressed by nouns or their equivalents. In this way "all horses have four feet" becomes "all horses are quadrupeds"; and, instead of "no horses eat flesh, or are carnivorous," we say, "no horses are flesh-eaters, or carnivora." Then, also, the predicate, which is commonly unquantified in the original proposition, must be given that quantity, whether universal or particular, which the nature of the sequence warrants. We must say — in thought, at least, — "all horses are some quadrupeds," and, "no horses are any flesh-eaters." After this quantification every ordinary affirmative statement asserts that all, or some, of the subject class, are identical with some, at least, of the predicate class; and every negative statement asserts that all, or some, of the subject class, are different from any — and, of course, from all — of the predicate class. Thereupon conversion follows on the principle of Identity. Because, so far as verbal thought goes, the converse of a thoroughly quantified dogmatic proposition presents essentially the same truth as the convertend.

But, as ordinary assertion aims only to characterize the subject, and does not quantify the predicate, the subject of the convertend loses its quantity after it becomes the predicate of the converse. "All horses are some quadrupeds," and "no horses are any carnivores," become, simply, "some quadrupeds are horses," and "no carnivores are horses." The ordinary purposes of predication do not require us to think and say "some quadrupeds are all (the) horses" and "no carnivores are any horses."

3. The converse both of the universal affirmative and of the particular affirmative proposition, is a particular affirmative; because ordinary affirmation only identifies the subject with a part of the predicate class. "All men are wise" and "some men are wise" alike yield "some wise beings are men." The converse of the universal, however, may be said to be a stronger assertion than that of the particular.

Quite in consistency with the foregoing, certain universal affirmatives are convertible into universal affirmatives; because they are statements which contain more than the ordinary universal affirmative. To indicate this, they have been symbolized by the vowel U, instead of A; and we are told that U may be converted into U. This class of propositions comprises all those in which the subject is an exact logical necessitant of the predicate. Accordingly, "all spirits are sentient" yields "all sentient beings are spirits"; provided the convertend be understood to teach that spirits have sentiency as a distinguishing attribute, or as a specific property. Definitions, also, belong to the class U, because they give the exact analytic equivalent of a thing. If "every moral law is a rule of right action," then "every rule of right action is a moral law."

4. Passing to negative propositions, it is evident that all of these, when quantified, assert that *all, or some, of the subject class, are not any of the predicate class.* In other words, they entirely exclude from the predicate class all or some of the subject class. On this account the particle "not," properly enough, has been called "the exclusive particle"; though this designation does not set forth its essential meaning.

The principle of Identity requires the converse of the universal negative to be an universal negative. Hence "no four-stomached animal is carnivorous," yields "no carnivore has four stomachs."

5. The particular negative is commonly said to be incapable of conversion; it is more exact to say that the negative proposition obtained by converting the particular negative has no predicative force. "Some colored men are not negroes," with quantified predicate, becomes "some colored men are not any negroes." The converse of this, "not any — or no — negroes are some colored men," is a true converse, yet *useless because of the indefinite "some."* For while negroes are not some colored men, they may be some other colored men. This converse does not enable us to say either that negroes are, or that negroes are not, colored; it does not characterize the subject either positively or negatively; therefore it fails of the essential end of predication.

But while the particular negative does not directly yield any usable converse, its contrapositive does; and, employing this, we convert the particular negatively indirectly. "Some colored men are not negroes," by contraposition (Chap. XV.), becomes "some colored men are men not negroes," a particular affirmative; from which we obtain "some men not negroes are colored."

Not only O, but A, may be converted by contraposition. The contrapositive of "all horses are quadrupeds," is "no horses are animals not quadrupeds"; from which we obtain the converse, "no animals not quadrupeds are horses." E also may be converted in this way. "No men are perfect," yields "all men are beings not perfect," and then "some beings not perfect are men." But this converse of E is a weak assertion, and is seldom used. The particular affirmative alone cannot be converted by contraposition; because its contrapositive is a particular negative. The contrapositive of "some men are happy" is "some men are not unhappy"; this has no usable converse.

In every contraposed proposition the original predicate conception is displaced by its contradictory, and, because this contradictory is generally a negative conception, contrapositive conversion has been called "conversion by infinitation;" that is, by the formation of negative conceptions. The original conception, however, is occasionally negative, and is then displaced by a positive conception. In this case the conversion does not depend on infinitation, but on the reverse process; which might be called re-finitation.

6. The ordinary conversion of *A* into *I* was styled by old logicians *"conversio per accidens";* which phrase signifies that, in the subject of the converse, the predicate conception of the convertend is not used simply, but with reference to some "accidental" addition. For, in saying conversely, "some animals are men," we do not mean that any animals, simply as such, are men, but only that certain animals which have characteristics *not necessarily connected with the nature of animals in general,* are men. The same idea is presented when *A* is said to be converted "by limitation."

Ordinarily, in this conversion of *A* "per accidens," or "by limitation," the subject of the converse loses its universality; it drops part of its force. The converse "some animals are men" means only that "some animals are (at least) some men." But occasionally, especially in syllogizing, the subject of the convertend, as predicate of the converse, retains its universality; so that we really assert "some animals are all the men." This mode of converting an universal or apodeictic proposition might be styled conversion "per differentiam," or, more exactly, conversion "by the retained-necessitant." For the subject of the convertend, as predicate of the converse, retains its necessitant value, and its "specific" membership in the class designated by the other term.

The conversion of *I* into *I*, and that of *E* into *E*, are commonly called "simple conversion," because the converse has the same quantity and quality with the convertend. This language should not be allowed to conceal the fact that these conversions depend upon entirely different laws, so far as their quantifications are concerned. *I* is converted on the same principle as *A*, that is "per accidens," or by limitation; *E* is converted on the principle of *negational exclusion*.

Let us now turn from the conversion of pure categorical propositions to that of modals. Modal conversion reveals the inner significance of dogmatic conversion; and explains the conversion of all illative propositions whatever.

7. At this point we meet the fact that *modal propositions often quantify their subjects* in the same way that dogmatic propositions do; and are compelled to enquire into the meaning of this quantification. We sometimes say, not simply "man must die; man cannot reach perfection," but "all men must die; no men can reach perfection"; sometimes, not simply "a professor of religion may be a hypocrite; a liquor-dealer may not be a bad man," but "some professors of religion may be hypocrites; some liquor-dealers may not be bad men." In short, necessary statements are occasionally given universal quantity, and contingent statements particular quantity.

Not only so; we sometimes "distribute" the subject in contingent statements, and employ "undistributed" subjects in necessary statements. We say, "all soldiers — or any soldier — may exhibit bravery; some soldiers must die in battle."

Let us consider, *first*, universal statements of necessity; *secondly*, particular statements of contingency; *thirdly*, universal statements of contingency; and *fourthly*, particular statements of necessity.

8. The universal necessary proposition differs from a simple general statement of necessity *only in being more explicit and emphatic*. "All men must die" means that "man, as such, must die." If man, simply as man, is mortal, then all men must die. But if man were necessarily mortal only when subjected to influences from which some of the race are free, we could not say that "all men must die," or that, absolutely speaking, "man must die." In that case we could only say, "some men must die," and, with regard to man as such, "man may die." Hence the universality of an apodeictic proposition shows that the statement is made unreservedly, and without mental qualification or limitation. It arises from, and is used to indicate, the absolute necessity of the statement. Therefore, also, when any proposition is given and accepted as a rule of necessary sequence and of demonstrative inference, the universality may be dispensed with.

For a similar reason the particular contingent proposition *need not be regarded as a necessary logical form.* "Some professors of religion *may* be hypocrites," as a general contingency, differs as to strength only from the assertion that "a professor of religion may be a hypocrite." Its meaning may be expressed without the "some" if we give the word "may" a questioning emphasis. It states that a professor of religion may be a hypocrite, but suggests also that the realization of this contingency is not to be expected under ordinary circumstances. It is consistent with the proposition that many professors of religion cannot be hypocrites. In short, a particular contingent proposition respecting a logical class sets forth such a weakened contingency as is suggestive of improbability. It should be recognized among the forms of modality. Yet the weak contingency which it embodies may be conceived and expressed without the quantification; we can therefore simplify our discussion — so long, at least, as it relates to mere contingency — by dispensing with this quantification.

9. *The universal contingent proposition,* as might be expected, has a force opposite to that of the particular. It *expresses a strong contingency;* especially when the universality is emphasized. "A professor of religion may be a hypocrite" is a contingent assertion applicable to every member of the class spoken of considered simply as a member of the class. This contingency is strengthened when we say, "*All* professors may be hypocrites." The first assertion would consist with the knowledge that some professors are not, and cannot be, hypocrites, though not of course, with such knowledge respecting any whose character is in question; the second assertion excludes such knowledge altogether. The same thoughts are expressed by contrasting "any professor may be a hypocrite" and "*every* professor may be a hypocrite." But should we omit the contrast and emphasize "*any,*" there would be no difference between these statements.

In short, there is no difference between an universal contingency and an unquantified contingency, if the latter be understood absolutely, or as excluding all knowledge of exceptions. The statement, "It may be that every liquor-dealer is a bad man," would express a strong contingency; for it would imply that one could not say that the rule has any exceptions.

10. Finally, propositions which set forth necessity (or impossibility) concerning an undistributed subject, *are really contingent assertions respecting the subject viewed simply.* "Some soldiers (that is, some of the logical class 'soldiers') must die in battle" expresses the contingent rule, "a soldier may die in battle."

The contingency thus expressed, however, is affected by two additions. *First,* we are informed that the antecedent of contingency, "a soldier in battle," is sometimes filled out so as to become an antecedent of necessity. This, also, is the essential thought expressed by the dogmatic proposition, "some soldiers die in battle." Hence,—as necessity conflicts with impossibility—we are informed that the contingent rule "a soldier may die in battle," cannot be supplanted by the apodeictic rule "a soldier cannot die in battle"; it is guarded against impossibility. In like manner, the principle of reasoning that "some soldiers cannot—or not all soldiers can—be killed in battle," and which, so far as it is a general contingency, is expressed by "a soldier may not be killed in battle," cannot be supplanted by "a soldier must be killed in battle"; it is guarded against necessity. This addition, whereby a contingency is guarded against impossibility, or against necessity, —or, in general terms, against a necessity of the opposite— is important, and cannot be neglected in the opposition and conversion of predications. The quantification employed in making the addition instrumentally determines and expresses the essential character of the proposition as regards modality. In short, the "some" of particular necessary propositions sets forth contingency in precisely the same way as the "some" of particular dogmatic propositions; and, in each case, the verbal thought should be distinguished and separated from its mental meaning.

Secondly, the "some" of the particular necessary proposition indicates that an appreciable proportion of the class "soldiers" are certain to die in battle, and, in so doing, brings before us, indirectly, the essential nature of contingency as distinguished from possibility in general. For, while contingency admits of various degrees, all contingency, even the weakest, is a strong possibility, circumscribed and determined by a necessity; and therefore such as justifies expectancy. It is possibility confined to a sphere in which only a limited number of consequents are possible.

As in every battle, or set of battles, some soldiers die while the rest survive, and there are thus as many events as there are soldiers, it follows that any soldier, taken at random, has so many chances to be killed and so many to live through the battle or the war. One of the deaths may be his or one of the survivals; and one out of the limited total of events *must* be his. Therefore the possibility of his being killed or not, is a circumscribed, or determined, possibility — a contingency. This contingency is further strengthened in case the "some" of the proposition respecting the class "soldiers," is conceived to be a considerable proportion of the "all."

If the ratio of the "some" to the "all" were fixed and given, a regular judgment of probability would take place. But that being unsettled, there is only a contingency, which approaches a probability without reaching it; or, if you please, a contingency which is an undetermined probability, while it is itself a determined possibility. Such is the significance of the particular assertion of necessity.

Yet not all contingency asserts that a certain event necessarily happens to a number of a logical class, and that any one in the class may be of that number. If some appreciable proportion of the balls in a box were red, there would be a contingency of corresponding strength that the ball *first* drawn out would be a red one; and were a thousand boxes similarly supplied, the contingency would be the same for box after box. This contingency would assume that one out of the limited number of balls in each box must appear; but it would not be based on any knowledge that red balls have appeared, or that *red balls must appear*, any number of times. Yet this contingency would be guarded, if we knew that there was nothing to prevent any ball from being drawn on any repetition of the trial; and it would become a definite probability, if we knew the exact number of balls of each color. Such contingency differs from that expressed by the "particular" proposition in its origin; but not in its nature, and as a ground and mode of judgment. And this being the case, it is plain that particular quantification is not necessarily connected with guarded contingency, but only with the origin of a certain form of it; which also it naturally expresses.

11. The foregoing analysis shows how quantification may strengthen and weaken, modify and define, modal assertion, while yet quantity is no proper part of modal thought, and has only a secondary and ministrant place in the expression of modality. The essential meaning of any modal proposition dispenses with quantification.

Such being the case, we proceed to the conversion of modals; beginning with the conversion of apodeictic assertions — that is, of propositions setting forth unqualified necessity and impossibility.

The converse of a necessity is a contingency. For, if a man must be a mortal, a mortal may be a man. To speak more accurately, it is a contingency guarded against impossibility. For if man, as such, must be a mortal, then a mortal, as such, may be a man, but *is certainly not incapable of being so.* This exclusion of impossibility, which results from the necessity asserted in the convertend, is often implied when we use the word "may" alone; and the exclusion of necessity may be implied in using "may not" alone. Then what "may be" cannot, on further information, be found impossible, and what "may not be" cannot be found necessary. The word "may" sometimes indicates an unstable possibility, in which case there is no exclusion of impossibility; it expresses the converse of necessity only when this exclusion is understood.

Moreover, the contingent converse of a necessity is *exclusive with reference to its subject*. The full statement of it is, not that "a mortal may be a man," but that "only a mortal may be a man." Commonly this exclusion, being unnecessary to the course of one's reasoning, is allowed to drop out of thought. But sometimes it is essential; and then it must be retained and recognized.

This full conversion of necessity might be distinguished as "differential conversion." It is expressed dogmatically when we say, "some mortals are all men." It is a peculiar case of limitative, or contingent, conversion. It might be named conversion with "the retained necessitant."

The contingency produced by the conversion of a necessity arises from the circumstance that the consequent of a necessary sequence conditions its antecedent: because, whenever we can assert that a condition of a first thing exists in a second thing, we can say that the first thing, so far forth, *may* be — or is possible. As the use of this law depends on the assertion of the consequent, it may be called *the principle of the asserted consequent*. The common rule is, that we cannot assert the antecedent because the consequent is asserted; this, however, means only that the antecedent cannot be asserted absolutely, or apodeictically; it can be asserted contingently. This principle is a part of the general theory of conditions.

12. The converse of an impossibility is an impossibility. "A horse cannot fly"; therefore "a flying animal cannot be a horse." This converse has the same modality as the convertend; there is no change either from necessity or from negation: hence the conversion is called "simple." The simplicity, however, is superficial; there is really a great change. In the convertend we reason *from an existing subject to a non-existing predicate* — from an existing horse to the non-existence of a flying animal in the horse. In the converse we no longer conceive of the original subject as existing, and of the original predicate as non-existent; but do just the reverse. We reason *from the existence of the predicate to the non-existence of the subject* — from the existing flying animal to the non-existence of a horse in it. This is a radical change.

The law governing this conversion is *the principle of the denied consequent*, that is, the principle which requires us to contradict the antecedent, if we contradict the consequent, of a necessary sequence. That such is the nature of the conversion will be evident, if we remember that the impossible and the necessary not to be, are the same. "Man cannot be perfect," means "man necessarily is not perfect." If now we contradict the consequent, "not perfect," by asserting "perfect," we must contradict the antecedent "man," in other words, *man as existing*. Therefore we say, "a perfect being necessarily is not — or cannot be — a man."

An impossibility is also convertible on the principle of the asserted consequent. "a man cannot be perfect" has "man" for antecedent, and "not perfect" for consequent. Asserting this consequent, we have, "a being not perfect may be a man." This converse, because of its contingency and of its negative subject, does not compare in value with the other, "a perfect being cannot be a man."

Necessity, also, may be converted on the principle of the denied consequent. "A war requires an army" has the converse, "where there is no army, there can be no war." So, since a plain must be extended, what is not extended cannot be a plain. This converse is apodeictic and absolute; and, notwithstanding its negative subject, is quite useful. It gives a test for the existence of any subject whose attributes or properties are known: if any of these do not exist, the subject cannot exist. It also furnishes the means of reducing a false statement of necessity to an absurdity. For if, in any instance of an assumed antecedent, its alleged necessary consequent can be shown to be wanting, this would lead to the contradiction and impossibility, that the antecedent known to exist does not exist.

13. The conversion of contingency, positive and negative, presents far more difficulty than that of necessity and impossibility. The laws of contingent conversion can be simply stated; but the intelligent use of them involves an understanding of the subtle compounded nature, and of the delicate variations, of contingent sequence. Besides, the ambiguity of language adds to the inherent obscurity of this subject. This especially attaches to the word "may," which sometimes denotes a naked, or bare, possibility, such as excites no expectancy; sometimes a clothed, or invested, possibility, which alone deserves the name of a contingency; and sometimes a specific mode of contingency: so that the meaning of this word must often be a matter for consideration.

The general rule for the conversion of a true contingency is that *we must follow the principle of the asserted, and not that of the denied, consequent.* "Man may be wise," of which the consequent is *wise as existing* in man, has the converse, "a wise being may be a man"; while "man may not be wise," of which the consequent is *wise as non-existent* in man, has the converse "a being not wise may be a man."

But should we apply the "denied consequent" to the first of these convertends, the result, "one not wise may not be a man" — though in a certain sense a correct inference — would have no predicative force. It sets forth a possibility which not only is unguarded against either necessity or impossibility, but is also unsupported by any ground for believing that the negative sequence contingently asserted has ever at any time been realized, or that it ever will be. A precisely parallel conversion would be, "a quadruped may be an elephant," therefore "an animal not an elephant may not be a quadruped" — an inference entirely nugatory, because, for aught that the convertend teaches, it may be true that all animals not elephants are quadrupeds, or that none of them are quadrupeds; and we are given no reason to suppose that any one of them ever has not been, or will not be, a quadruped.

In like manner, applying "the denied consequent" to the convertend "man may not be wise," we have "a wise being may not be a man," a possibility wholly unprotected, indeterminate, and without predicative force. Because, for aught that is given in the convertend, it may be true that a wise being must be a man, or that he cannot be a man, and we have no reason to believe that any wise being ever was not, or will not be, a man. A parallel conversion to this would be, "a quadruped may not be an elephant," therefore, "an elephant may not be a quadruped."

14. The conversion of contingent modals is closely related to that of particular dogmatic propositions. These latter, internally and essentially, are simply the most common and important forms of contingent sequence. They are contingencies guarded against the necessity of the opposite. Hence the rules for their conversion can be explained by the laws of modal con-

version, even while they exhibit an apparent contrast to these laws. For contingent propositions, positive and negative, are converted according to one principle (the asserted consequent) which applies to both alike; while particular dogmatic propositions are converted by two rules. *I*— the particular affirmative — is converted by "limitation," and *O*— the particular negative — by "contraposition." Thus "some men are wise," converted by limitation, yields "some wise beings are men"; and "some men are not wise," converted by contraposition, yields "some beings not wise are men."

This contrast is a result of that form of thought which *negative propositions with a substantal predicate* naturally assume, and which is especially observable in the pure, or dogmatic, negative. In all such propositions we aim to assert *the non-existence of an identity*. But, ordinarily, the mind, clinging to positive conception, instead of asserting that non-existence immediately, first conceives, though without assertion, of the identity as existing, and then denies its existence. This mode of conception resembles that according to which negative necessity becomes impossibility. Contraposition destroys the indirectness of such assertion by immediately attaching the thought of non-existence to the predicate, and then substantializes the negative conception thus produced. After this change, the predicate, "not wise being," truly sets forth the contingent negative consequent; which the original predicate, "wise being," did not. Thereupon the contraposed proposition, as consisting of antecedent and consequent, is converted on the same principle as the particular affirmative, that is, "by limitation"; which is equivalent to "the asserted consequent," as applied to guarded contingencies.

CHAPTER XXI.

CONTINGENCY AND ITS CONVERSION.

A SUPPLEMENTARY CHAPTER.

1. Contingency distinguished from possibility. 2. Possibility defined. 3. Contingency, a circumscribed possibility. 4. Either empirical or mathematical. 5. Involves an opposite possibility, but not an opposite contingency. 6. Is either guarded (i.e. half-guarded) or unguarded. When double, may be doubly guarded. 7. When combined with a prior sequence, produces a contingency unguarded, *i.e.* unassured, against a necessity of the opposite. 8. Unguarded mathematical (or intuitional) contingency. 9. Embedded contingency, is contingency only in an improper sense. 10. Possibility is converted only by the asserted-consequent; and according to (*a*) the law of contained-conditions, and (*b*) the law of the unascertained-necessitant. 11. The converse by the denied-consequent has no force, or value. 12. The conversion of contingency. Violently rejects the denied-consequent. 13. Is effected by the asserted-consequent. 14. Assumes a numerical limitation of the predicate of the convertend. 15. Not ordinarily, nor necessarily, double. 16. The converse of an encouraging — or guarded affirmative — contingency, is an encouraging contingency with a positive subject. 17. That of a discouraging — or guarded negative — contingency is an encouraging contingency with a negative subject. 18. These unite in the double converse of a double guarded contingency. 19. The conversa of unstable contingencies are unstable. 20. The conversion of improper — or embedded — contingency follows that of the necessity in which it lies. 21. A scheme of symbols.

1. THE perplexity which has hitherto obscured the exposition of modal assertion and reasoning has arisen principally in connection with contingent propositions. The nature of apodeictic statements is easily understood. He who would cut plain roads through the labyrinth of modality must set forth the forms and laws of contingent predication. We have already attempted this; but an additional and somewhat independent discussion will be found useful.

Contingency, as a ground of inference, is that mode of possibility which excites expectancy; it may be distinguished from simple possibility by the conditions under which it is produced.

2. Possibility, in the widest sense, is the *consistency* of the existence of a thing with given surroundings. What is consistent with given circumstances is not impossible in those circumstances. This wide possibility shows that the question as to the reality of a thing is not absolutely absurd.

Ordinary logical possibility, however, is more than mere consistency, or non-repugnancy. It is the *compatibility*, or *congruity*, of the existence of a thing, with given circumstances. Hence it is suggestive of the question of reality, even while it may suggest no answer to this question. For when we say that B is possible because A exists, meaning that the existence of A involves the existential compatibility of B with the circumstances assumed with A, the question arises, "Does B exist?"

The compatibility, or congruity, of B with given surroundings, rests on the fact that A, as one of those circumstances, contains some, at least, of the necessary conditions of B. A broken line composed of three straight lines on a plane renders a triangle possible because it contains three conditions of a triangle. But, in order to a suggestive sequence of possibility, the conditions contained in the antecedent must be such as specially connect themselves with B, the consequent. If they are of a very general character, they will not imply the possibility of B specifically. It would not be a suggestive sequence to say that space renders a line, or a triangle, or a field, or a house, possible. Such judgments are metaphysical rather than logical. But the specific judgments, "a *line* may be straight," "a *triangle* may be scalene," "a *house* may be of four stories," might prove suggestive and useful.

Whenever an antecedent of possibility is perceived to contain such a combination of conditions as necessitates the consequent, it becomes an antecedent of necessity, as well as of possibility. Ordinarily, however, the antecedent of possibility either does not have a necessitant force or is not perceived to have it; so that the question of reality is left undetermined, and even untouched. For, as already said, possibility *per se* suggests no answer to this question. Its judgments result in the harmonious construction of thought, but are only negatively helpful towards the ascertainment of truth.

3. Possibility excites expectancy only when it is strengthened into contingency. For contingency is a ground for believing, not simply that a thing is abstractly possible, but that it actually may be true.

This latter mode of sequence is often asserted in an unreasoned way, or, rather, by an intuitive and practical exercise of the reason. Perceiving that one certain kind of fact or event is occasionally followed by another, we not only associate the latter with the former, but regard it as contingently connected with the other, and to be looked for, with more or less expectation, whenever the other occurs. But when we reflect on such a judgment, so as to make it understandingly and place it on a foundation, we find that the possibility — the contingency — which it asserts, is confined to a sphere in which only a limited number of events are possible, and in which one of these events must take place. Contingency, therefore, is a *circumscribed*, or *determined*, possibility.

The essential nature of contingency may be understood from the two following illustrations. Should we know that some of a limited number, say of a hundred, balls are red, without knowing how many, it would be a contingency to any ball taken at random to be red. The fact that the ball is one of the hundred is the antecedent of contingency; and it has a hundred possible consequents, of which an undetermined proportion favor the appearance of a red ball. It would be a variation of this illustration if there were an indefinite aggregation of balls, composed of many equal sets in each of which there were some red balls; in this case also it would be contingent to a ball taken at random to be red.

Again, if we knew that some snakes are venomous, without knowing what proportion, it would be contingent to any snake, taken at random, to be venomous. The antecedent here is membership in a class of things which sometimes have a certain character — in other words, the possession of that snake nature, which sometimes is venomous.

4. So far as the foregoing judgments assert contingent sequence they both arise in the same way: they both make an indeterminate use of the tychologic principle — "the ratio of the chances." But *they differ as to the process by which each forms and accepts the conception of favoring chances.* That snakes are sometimes venomous has been ascertained from observation, and is the ground for an homologic inference. An indefinite proportion of all snakes hitherto seen having been found venomous, this may be asserted concerning snakes not yet seen: so we say that some of the whole logical class are venomous. This justifies the general judgment "a snake may be venomous." The contingency, thus expressed, is perceived only after a previous perception of actual sequences, and, with reference to this, may be named *inductive*, or *empirical*, contingency. But the contingency that any ball of the hundred or more may be red, rests on our immediate knowledge respecting a set, or an aggregation, of balls, that some of them are red, and has no connection with any previous experience. It does not assume that, in some previous trials, the ball chosen at random has turned out red.

Contingent judgments of this latter formation are less frequent than those based on the observation of past sequences, yet they illustrate better the essential principle of contingency; for they make no addition to it. This form of contingency is that assumed by mathematicians, and may be distinguished as *intuitive*, or *mathematical*. J. S. Mill and the Associationalists teach that all contingent judgment is empirical, or based on observation of the past; their doctrine gives no satisfactory account of mathematical contingency.

Such is the nature of contingency, not as a general, but as a specific, mode of logical sequence. It lies between possibility and probability, and is more determinate than the former, and less determinate than the latter.

5. Two characteristics of contingency are closely connected with its nature. In the first place, such sequence is always accompanied with a "possibility of the opposite." The "opposite"

here means the contradictory of that which is contingently asserted. When we judge that the ball selected at random may be red, or that the snake met accidentally may be venomous, it is also felt that the ball may not be red, and that the snake may not be venomous. This doubleness arises because the antecedent of possibility both assures us that some conditions of the consequent exist, and leaves us in doubt whether or not others do; it therefore justifies both a positive and a negative sequence in possibility. But *contingency is double only when both sides of the possibility are supported by known facts or instances.* Consequently we cannot say that contingency is always double, but only that it is always accompanied by a possibility of the opposite. Knowing that some snakes are venomous, but not whether any snakes are not venomous, we can assert a negative possibility, but not a negative contingency, concerning snakes. This negative possibility accompanies the positive contingency, like a shadow. So also a positive possibility accompanies a negative contingency. These possibilities differ from the contingencies which they accompany in that they are not grounds of expectancy. Because, for aught we know, it may be true that they never have been — or that they never can be — realized in any case.

It may be said that the opposite of the contingency asserted is supported by any chances that remain after those favoring the assertion have been subtracted from the total number; and that, therefore, the opposite of a contingency is also necessarily a contingency. This, however, is not so. The *denial* of the opposite of the asserted contingency is supported by the same chances which support the original assertion, and is also a contingency. But the *assertion* of that opposite is not really supported by any chances at all. For, as the remainder above mentioned may be either some chances or none, we have no right to depend upon any. In short, the opposite of the contingency asserted, being wholly unsupported by facts or instances, is only a naked possibility.

6. The second characteristic of contingent sequence is that it may be either guarded or unguarded. Naturally and primarily positive contingency is guarded against impossibility, but

not against necessity; while negative contingency is guarded against necessity, but not against impossibility. Each of these, therefore, may be termed half-guarded, or — more simply — guarded, contingency; and, as we shall see, each of them may become unguarded. Knowing simply that some snakes are venomous we have the guarded contingency, "a snake may be venomous": knowing simply that some snakes are not venomous, we have the guarded contingency, "a snake may not be venomous." These contingencies may combine in the double sequence, "a snake may, and may not, be venomous," which appeals to both positive and negative instances. This, as guarded against both impossibility and necessity, may be described as doubly guarded.

7. Contingency loses its guarded character if it be not immediately based on facts, but inferred from the combination of a contingency with a prior sequence. Knowing that "a lion is a quadruped," and that "a quadruped may be a carnivore," we say, "A lion may be a carnivore." Also knowing that "a lion is a quadruped," and that "a quadruped may not be a carnivore," we say, "A lion may not be a carnivore." Further information displaces both of these deduced contingencies by a necessity. In like manner the similarly inferred contingencies, "an ox may be a carnivore" and "an ox may not be a carnivore," give place to an impossibility. Again, some reptiles being snakes and some snakes venomous, we say, "A reptile may be venomous"; and, for like reasons, "A reptile may not be venomous." Here are two unguarded contingencies; additional knowledge renders each as guarded as those from which it has been inferred. An exception to the rule now explained will be noticed hereafter.

Unguarded contingency may be single or double, according as the contingency from which it is deduced is single or double. In the above illustrations, if we unite the opposite single assertions, we can say, "A lion may, and may not, be a carnivore," "An ox may, and may not, be a carnivore," and, "A reptile may, and may not, be venomous." But it is not so important to distinguish between the single and the double mode of unguarded contingency as it is to distinguish between that guarded contingency which is double, being both positive and negative, and that guarded, or half-guarded, contingency, which is single, being either positive or negative. Ordinarily contingency is single, and guarded only on one side.

8. Mathematical unguarded contingency may be illustrated thus: let there be 100 balls of ivory, all of these being white; 100 of wood, some of these being red; and let all the balls be placed in one collection. Then if one knew not that all the ivory balls are white, but only that (*a*) they are all among the 200, and that (*b*) some of the 200 are red, it would be a contingency to an ivory ball to be red. Or, if he knew only that (*a*) all the ivory balls are among the 200, and that (*b*) some of the 200 are white, it would be contingent to an ivory ball to be white. Investigation would displace either of these contingencies by the certainty that every ivory ball is white. But if the ivory balls were some white and some red, and this should appear on investigation, then the contingent judgments respecting the color of any ivory ball (taken at random) would become guarded, and stable. For unguarded contingency may be termed unstable; guarded (or doubly guarded) contingency, stable; and the half-guarded, half-stable.

9. In addition to the foregoing modes of contingency, we must mention that fixed, or embedded, possibility, which may sometimes be called contingency; and which is that compatibility of the existence, or of the non-existence, of a thing with given circumstances, which may be inferred from necessity, or from impossibility. This mode of sequence is possibility or contingency only in an improper sense; for it excludes the possibility of the opposite; but it has a place in logic.

Comparing with each other the different modes of contingency proper, we find that guarded contingency is the most developed and complete sequence; half-guarded contingency is the most frequently used in reasonings; and unguarded contingency is the purest, but also the weakest and least determinate, mode of contingent sequence.

The Conversion of Contingency.

10. The general rule for the conversion of possibility and of contingency is that *either may be converted by the asserted-consequent, but not by the denied-consequent.* To understand this rule we must discuss first the conversion of possibility, and then that of contingency.

With respect to possibility let us first show that the asserted-consequent yields a logical converse, and then that the denied-consequent does not do so.

The following specific formulas exhibit the conversion of possibility by the asserted-consequent:

(1) If the existence of one thing (A) render possible the existence of another thing (B), then will the existence of B render possible the existence of A.

(2) If the existence of A render possible the non-existence of B, then will the non-existence of B render possible the existence of A.

(3) If the non-existence of A render possible the existence of B, then will the existence of B render possible the non-existence of A. And

(4) If the non-existence of A render possible the non-existence of B, then will the non-existence of B render possible the non-existence of A.

The non-existence mentioned in these formulas always relates to, and is included in, a case in which something is non-existent; it is not non-existence *per se*. Non-entity, of itself, is never either antecedent or consequent; but cases occur in which the non-existence of one thing makes something else possible to be, or possible not to be. Those antecedents which assume non-existence, are cases of existence modified by the non-existence of some element which might have been present. Therefore, for the sake of simplicity, we may disregard the difference between positive and negative antecedents, and retain only the first two of the foregoing rules; after which these may be combined in the one rule that "if the existence of A render possible the existence, or the non-existence, of B, then will the existence, or the non-existence, of B render possible the existence of A." That is to say, any antecedent of possibility may be made the consequent of its own consequent. In other words, every sequence in possibility may be converted by "the asserted consequent."

This formula may be justified, first, with reference to affirmative possibility, and then with reference to negative possibility. The principle which gives vitality to affirmative possibility may be called *"the law of contained conditions"*; and the conversion of this mode of possibility by the asserted consequent follows upon the fact that the law of contained conditions has a reciprocal action. For the antecedent of possibility always contains a condition, or conditions, of the consequent; and the consequent, a condition, or conditions, of the antecedent. Let A be antecedent of possibility to B, because A involves c, which is a condition of B. Then, first, c is a condition of A, as being involved in A; and secondly, c is involved in B, as being a condition of B. This being so, B, as involving c, which is a condition of A, may be antecedent of possibility to A. Take the sequence, "man (A) may be wise (B)." Here wisdom is possible because man has intellect (c), which is a condition of wisdom. But intellect is a condition of "man," as being a necessary part of him; and it is involved in wisdom as being a condition of wisdom. Therefore, conversely, "a wise being may be a man." A coin may be a piece of silver, and a piece of silver a coin, because each of these involves "valuable metal." A long walk and a wide plain render each other logically possible, because each involves the element of "distance."

The principle which gives vitality to negative sequences in possibility is a corollary, or concomitant, of "the law of contained conditions"; it may be named *"the law of the unascertained necessitant"*; and this principle, like that which it accompanies, has a reciprocal action. In every assertion of possibility proper, while knowing that some conditions of an entity exist, we are ignorant whether such and so many exist as constitute a logical, or necessitating, condition. We may know that the antecedent, considered *per se*, does not contain a logical condition (or necessitant), or we may be ignorant whether it does or not; in the one case we assert a settled or stable, in the other, an unstable, possibility of non-existence; in either case we assert the possible non-existence of B, *because A (either as known, or so far as known) does not contain a necessitant of B*. But the non-existence of B, though involving the non-existence of any necessitant of B, and of any antecedent containing that necessitant, *is consistent with what does not contain the necessitant*. Therefore the non-existence of B·is consistent with the existence of A: "man may be wise" yields, first, "man may not be wise," and then the converse possibility, "a being not wise may be a man." Therefore, though there is silver there may be no coin, and though there be no coin there may be silver.

Ordinarily A is *known not* to contain a necessitant of B; so that the contingency, "A is possibly not B — man is possibly not wise" is guarded against necessity. In this case the converse, "Not-B is possibly A — not-wise is possibly man" is guarded against impossibility. But should A be only *not known* to contain a necessitant, the convertend would not be guarded against necessity, nor the converse against impossibility. Knowing that a carnivore may not be a quadruped, and that a lion is a carnivore, we may say, "A lion may not be a quadruped." Further knowledge will displace this by a necessity: and the converse, "a non-quadruped may be a lion" will be displaced by impossibility. But commonly the convertend is understood as guarded, so that the converse, also, is guarded. So much for "the asserted-consequent."

11. We are now prepared to ask whether possible sequence can be converted by "the denied-consequent," as well as by "the asserted-consequent." This point may be discussed as a question respecting the validity of two formulas, if, as before, we neglect the distinction between positive and negative antecedents, and so reduce four formulas to two. These are:

1. If the existence of A render possible the existence of B, then the non-existence of B will render possible the non-existence of A.

2. If the existence of A render possible the non-existence of B, then will the existence of B render possible the non-existence of A. Expressed categorically, these conversions are,

(1) *A* is possibly *B*; therefore, what is not *B* is possibly not *A*; and

(2) *A* is possibly not *B*; therefore, *B* is possibly not *A*. Coin is possibly silver; therefore, what is not silver is possibly not coin. — Coin is possibly not silver; therefore, silver is possibly not coin. In these proposed inferences, as the method of "the denied-consequent" requires, the contradictory of the consequent is used for antecedent and the contradictory of the antecedent for consequent.

The conversion of a *positive* sequence is attempted, in this way, on the principle that *the absence, or contradiction, of the antecedent renders the absence of its consequent possible* — that is, shows it to be possible. For the absence of the antecedent puts us in doubt whether even those conditions of the consequent respecting which the antecedent would give assurance, are present or not; inasmuch as, if they are, it must be in some other antecedent. Beyond question the law of "the unascertained necessitant" applies here in a very literal way; and so we say, first, "*A* is possibly *B*"; then, "What is not *A* is possibly not *B*"; after which, using "the asserted-consequent," as with any negative possibility, we obtain the converse, "what is not *B* is possibly not *A*." "Coin is possibly silver — what is not coin is possibly not silver — what is not silver is possibly not coin."

This sequence is correct; and *yet it is entirely nugatory and useless.* Though supported by the fact that the denial of an antecedent of possibility leaves no ground for conjecturing that the consequent exists, so that, until we learn more, we can say that, so far as we know, the consequent may not exist; it is open to two objections. The fatal objection is that the secondary, or intermediate, convertend, from which the converse is immediately produced, is without convictional value, because *it is founded purely on "the unascertained necessitant"; which principle is useless except as a concomitant principle.* All logical force disappears when we form that secondary convertend, by using contradictory conceptions. Hence no connection of congruity or compatibility is perceivable in the converse, between antecedent and consequent.

Then, secondly, while the original convertend may be a guarded possibility, *the converse is unguarded.* We correctly say, "A quadruped may be a lion — a non-quadruped may not be a lion — a non-lion may not be a quadruped." But, notwithstanding all this, it might be true that every "non-lion" is a quadruped.

The conversion of a *negative* sequence by "the denied-consequent" may be attempted as follows. "*A* is possibly not *B*. — Not-*A* is possibly *B*. — *B* is possibly not *A*." Coin is possibly not silver; what is not coin is possibly silver; silver is possibly not coin. Here, as before, the operation of "the denied-consequent" is equivalent to that of the asserted-consequent after contradictory conceptions have been employed.

This conversion, like that just considered, is without convictional force. In saying, "What is not coin may be silver," because "coin may not be silver," we base a sequence *simply on the removal of an antecedent of possible non-existence;* we assert a possibility because we have no reason either for or against it, except the removal of that antecedent. Such an assertion is entirely indeterminate; and so is the converse of it, "silver may not be a coin." Moreover, while the original possibility may be — and commonly is — guarded, this converse is unguarded. Should we say, "A quadruped may not be a lion; therefore a lion may not be a quadruped"; this converse will be displaced, on further knowledge, by a necessity.

12. We pass, now, from the conversion of possibility in general to that of contingency. By this we mean *the inference of a converse contingency from a convertend contingency;* for to infer a possibility conversely from a contingency, would only be a conversion in possibility, and not, distinctively, a conversion in contingency.

While the conversion of contingency follows the same rule as that of possibility in general, it has some noteworthy peculiarities. In the first place, *the method of the denied-consequent is more violently rejected by contingency than by possibility.* This method leads to a formal but useless converse in possibility, but produces no converse whatever in contingency. The reason for this is that the facts or instances which sustain the original contingency, do not support the proposed converse contingency.

The contingency, "a snake may be venomous," rests on the fact that "some snakes, at least, are venomous." This fact yields no support to the converse, that "an animal not venomous may not be a snake"; it is not a fact relating to such animals. So also the contingency, "a snake may not be venomous," rests on the fact that "some snakes are not venomous"; and this does not support any converse contingency respecting "venomous" animals. In short, the conversion of contingency by the denied-consequent, results only in a useless indeterminate possibility.

13. On the other hand, the asserted-consequent produces a true conversion; because the same facts which support the convertend, support the converse also. The same instances justify the contingency, "a snake may be venomous," and the contingency, "a venomous animal may be a snake." In like manner, the contingencies, "a snake may not be venomous," and "a non-venomous animal may be a snake," are supported by the same instances.

Mathematical contingency, equally with the empirical, is convertible by the asserted-consequent. Let some balls in a collection of one hundred be red. Then it is contingent to any ball, selected at random, to be red, and to any red ball to be the one so taken. Or, if some of the balls be not red, it is contingent to any ball selected at random, not to be red, and to any ball not red to be so selected. Convertend and converse originate together, and are supported by the very same facts.

14. But, in this connection, it should be remarked that converse does not follow convertend so absolutely — so perfectly as a matter of course — in contingency as in possibility. A converse contingency, unlike a converse possibility, depends on a limitation which, ordinarily and naturally, attaches to the predicate of the convertend, yet which is not necessarily inherent in it. For the class or set of things, to which the subject of the converse refers must be numerically limited in order that some indefinite proportion — or ratio of chances — may be assumed between the "some" and the "all." Without this limitation, at least in our first apprehension of the converse contingency, no basis of expectancy could be formed. But the

class thus numerically limited is the same as that to which the predicate of the convertend refers. In converting "a snake may be venomous," we assume that the venomous animals which are snakes belong to a class "venomous," and constitute an appreciable proportion of that class. In the converse of the negative contingency a similar ratio is assumed between the "non-venomous," which are snakes, and the whole class "non-venomous." So, in converting the mathematical contingencies, the "red balls" and the "balls not red" are thought of as belonging to the collection in the box; and not as being any red balls whatever, or any balls not red.

This numerical limitation somewhat resembles "quantification" of the predicate, but is quite another thing; for it is not exclusively related to a logical class.

15. Another difference between contingency and possibility is that the *conversion of possibility always admits of a doubleness, while this is not the case with contingency.* Every antecedent of possibility proper justifies both a positive and a negative consequent. Hence every positive sequence in possibility is accompanied by a negative sequence, and every negative, by a positive. This being so, the converse of a positive possibility is accompanied by the converse of the negative, and the converse of the negative by that of the positive. Therefore "a man may be wise," as a possibility, has the double converse, "a wise being may be a man," and "a being not wise may be a man." And, in the same way, both these conversa may be inferred from the negative possibility, "a man may not be wise." But the positive contingency, "a man may be wise," justifies only its own single converse; and the negative contingency, "a man may not be wise," only its own single converse. Neither of these contingencies can claim the converse of the other along with its own; because the facts supporting it justify only one converse contingency.

When a positive and a negative contingency are united so as to form a double contingency, the converse of the double contingency is also double; but this is not because each contingency warrants the converse of the other, but only because each is followed by its own.

16. In addition to the supreme law for the conversion of contingency some subordinate rules claim attention. These pertain to the different modes of contingency according as it is proper or improper, guarded or unguarded. In discussing them we need not continue to contrast possibility and contingency; for we must employ principles freely applicable to both.

The most common modes of contingency are that affirmative sequence which is guarded against impossibility, and which has been styled "encouraging," and that negative sequence which is guarded against necessity, and which we have named "discouraging." These correspond with the half-guarded modes of possibility, positive and negative; and are based on these possibilities. Both may be styled "guarded" in the sense that each is guarded against a necessity of the opposite.

The converse of an encouraging contingency is *an encouraging contingency with a positive subject.* If "a man may be wise," then "a wise being may be a man." The same instances support both these contingencies, and guard both against impossibility. The strength of the converse depends on the ratio of the men who are wise to the whole class "wise"; and varies with our estimate of that ratio.

17. The converse of a discouraging contingency *is an encouraging contingency with a negative subject.* If "a man may not be wise," then "a being not wise may be a man." The same facts justify both these contingencies. The converse is guarded against impossibility; because, by reason of the law of Contradiction, if any subject — A — be not a given predicate — B, then A is something which is not B. Therefore, on the same basis of fact, we say, "A man may not be wise — A man may be a being not wise — A being not wise may be a man." This last assertion is an encouraging contingency.

A discouraging contingency does not yield a discouraging converse, because this would involve "the denied consequent."

18. Encouraging and discouraging contingency are the two modes of half-stable contingency. Stable, or double-guarded, contingency is the compound from their conjunction. Accordingly *the converse of stable contingency is two-fold, and includes the converse of each of the constituent parts.* Knowing that some men

are wise and some men not wise, we have the stable contingency, "a man may, and may not, be wise," with the double converse, "a wise being may be a man, and a being not wise may be a man"; each of these assertions being a half-stable encouraging contingency.

But we cannot say, conversely, "A wise being may, and may not, be a man," because the negative part of this converse would involve the denied-consequent.

19. *The converse of an unstable contingency is an unstable contingency.* The original assertion being only mediately and contingently supported by facts, this must be the case with the inferred proposition also. Knowing simply that "some carnivores are quadrupeds, and some quadrupeds lions," we say, "A carnivore may be a lion." This is an unstable contingency; further information might show that a carnivore cannot be a lion, or that it must be a lion. For the same reason the converse, "a lion may be a carnivore," is unstable; and further knowledge will show that lions are necessarily carnivorous.

Again, knowing merely that "all oxen are quadrupeds and that some quadrupeds are carnivores," we have the unstable contingency, "an ox may be a carnivore," and its converse, "a carnivore may be an ox." Further information displaces both convertend and converse by an impossibility.

Once more, knowing only that "some mammals are quadrupeds and some quadrupeds are carnivores," we have the contingency, "a mammal may be a carnivore," and its converse, "a carnivore may be a mammal." Both are unstable; further knowledge renders both stable. For it is neither necessary nor impossible that a carnivore should be a mammal, or that a mammal should be a carnivore.

The foregoing contingencies are single. Should we say, "An ox may, and may not, be a carnivore," because "a quadruped may, and may not, be a carnivore," we should assert a double unstable contingency; and its double converse, "a carnivore — as also a non-carnivore — may be an ox," would consist of two unstable assertions.

20. The foregoing laws of conversion are those of contingency proper in its various modes, and do not control fixed, or

embedded, contingency. This has the peculiarity that it may be converted either by the asserted consequent or by the denied consequent — by the former because it participates in the nature of contingency (though not a true contingency); by the latter because it shares in the relations of necessity. The possibility, "man may be mortal, because man must die," yields not only "a mortal may be a man," but also "what is not a mortal may not be a man." For this latter contingency is embedded in the converse, "what is not a mortal cannot be a man"; which is obtained by the denied consequent from the original underlying necessity. In like manner, the possibility, "man may not be perfect, because man cannot be perfect," yields, not only "a being not perfect may be a man," but also "a perfect being may not be a man." This is embedded in the converse of the underlying impossibility.

Such being the case, it is plain that the converse of a fixed contingency by the denied consequent is another fixed contingency. But this is not the result when the asserted consequent is used. Then the converse of a fixed contingency is the same as the ordinary converse of necessity (Chap. XX.). More specifically, the converse of a positive fixed contingency is an encouraging contingency with a positive subject, while that of a negative fixed contingency is an encouraging contingency with a negative subject. Thus the embedded contingency, "man may be mortal," yields the encouraging contingency, "a mortal may be a man": and, in like manner, "man may not be perfect" yields "an imperfect being may be a man."

21. Some advantage might result if the various modes of possibility and contingency were indicated by symbols. In particular the student might construct for himself a useful scheme of those oppositions and conversions in which possibility and contingency are concerned. To this end we make the following suggestions. Let the small Greek vowels ι and ο indicate the positive and negative modes of unguarded, or unstable, possibility; that being the purest form of possibility proper. Let possibility as guarded against impossibility be marked by the grave accent, thus, ὶ and ὸ; as guarded against

necessity, by the acute accent, thus, $ί$ and $ό$; and as guarded against both impossibility and necessity, by the circumflex accents, thus, $ῖ$ and $ô$. In possibility proper $ι$ and $ο$ always accompany each other. So, also, in the modes of guarded possibility, do $ὶ$ and $ὸ$; $ί$ and $ό$; and $ῖ$ and $ô$. The two modes of embedded possibility might be indicated by the same letters enclosed in parenthesis — ($ι$) and ($ο$). These do not accompany each other.

The different modes of contingency might be symbolized by circumscribing with a circle those proper possibilities on which contingencies are based. Thus, ⓘ and ⓞ may indicate single unstable contingencies; $\overline{ⓘ+ⓞ}$ a double unstable contingency; ⓘ̀ and ⓞ́ are half-guarded contingencies; $\overline{ⓘ̂+ⓞ̂}$ is stable contingency.

But, for the sake of simplicity, let the diphthongs $ει$ and $ου$ take the place of the circumscribed vowels. Then $ει$ and $ου$ and $\overline{ει+ου}$ indicate the forms of unstable contingency; $εί$ and $ού$ the half-guarded contingencies; and $\overline{εῖ+οῦ}$ the guarded; that is, the doubly guarded.

Every single contingency *embraces* a corresponding possibility and *is attended* by a possibility of the opposite; but not by a contingency of the opposite. Thus $εί$ and $ού$ do not necessarily accompany each other; but $εί$ embraces $ὶ$, and is attended by $ὸ$, and $ού$ embraces $ό$, and is attended by $ί$.

So, in unstable contingency, $ει$ involves $ι$ and $ο$, but not $ου$; and $ου$, $ο$ and $ι$, but not $ει$.

The foregoing discussions show that the logician is compelled to employ the conception of contingency more specifically and definitely in connection with the conversion, than in connection with the opposition, of predications. We account for this, because opposition deals with given propositions, while conversion is the formation of a new statement; and because, while contingency and possibility, by reason of their common nature, may be used in similar dialectic oppositions, their conversions differ by reason of the specific differences belonging to them as modes of sequence.

CHAPTER XXII.

SYLLOGISMS.

1. Syllogisms defined. 2. The syllogism-proper. 3. Relational syllogisms: (*a*) immediate, (*b*) mediate. 4. Homologic syllogisms: (*a*) paradigmatic, (*b*) principiative, (*c*) applicative. 5. Hypothetical syllogisms. 6. The consequent-consequent is the first and supreme law of syllogisms-proper. 7. The principle of the separating-consequents. 8. The principle of the common-antecedent. 9. The principles of syllogistic reciprocation. 10. These are less independent in their operation than the other laws. 11. The three propositions, and the three terms, of the syllogism. 12. To analyze a syllogism, begin with the conclusion. 13. The four "figures." The order of the propositions. 14. Syllogistic moods.

1. "A SYLLOGISM," says Aristotle, "is a statement in which, certain things being laid down, something else, different from the premises, necessarily follows in consequence of the premises" ("Topics," I. 1). The "things laid down," or "premises," are propositions known, or assumed as true; and the "something else" is a proposition, either apodeictic or problematic, necessarily believed in consequence of the premises; but the main teaching of the definition is that syllogistic inference arises from more than one premise. This, indeed, is the essential meaning of the noun συλλογισμός as derived from the verb συλλογίζεσθαι. For συλλογίζεσθαι (συλλέγειν) indicates the gathering, or collecting, of certain elements from given premises, and putting them together, so as to form a conclusion. Yet the plurality of premises does not involve a plurality of antecedents; the combination of the premises is necessary to constitute one antecedent.

Of late years any formal inference, even though it should have only one premise, has been called a syllogism. For example, "This is an action; therefore there is an agent, — This is an event; therefore there is a cause, — Air is a substance; therefore it occupies space,—All trees spring from seeds; therefore these trees have done so,"—have been classed as immediate syllogisms. Let us now restrict the term to inferences of more than one premise.

Moreover, as every such inference, when formally expressed, either naturally or necessarily uses two premises, let us mean by syllogism the statement of a double-grounded inference. All the syllogisms of Aristotle have this character. Nay; the forms and rules of syllogizing given by Aristotle, and which chiefly call for study, do not apply to every kind of double-grounded inference, but only to one important mode of it; which, therefore, may be distinguished as the syllogism proper, the syllogism *par excellence*. In the following discussion we shall explain the radical nature of the true Aristotelian syllogism, after first describing some other forms of double-grounded inference.

2. Syllogisms proper are *inferences in which from two general illative propositions a third general illative proposition is deduced;* improper syllogisms are inferences in which from two propositions, one of which at least need not be a general illative proposition, a third proposition is deduced.

Dividing improper syllogisms into three classes, according to their formative laws, we shall have, in all, four classes of syllogisms. These may be named (1) the relational, (2) the homologic, (3) the hypothetical, or translative, and (4) the catenate. Syllogizing proper is catenated inference; because, by means of it, we form chains of abstract reasoning.

3. Relational syllogisms are scarcely worthy of the name. They are orthologic sequences made according to different specific laws, and are distinguished from other sequences of that class only by the complexity of their antecedents. They may be subdivided into the (*a*) immediate and the (*b*) mediate; though these designations are somewhat ambiguous and inadequate. The immediate may be illustrated as follows:

> This is a line;
> And it is straight; therefore
> It is the shortest possible between its terminal points.

These lines are straight;
And they are parallel; therefore
They will continue parallel, however prolonged.

A and B are respectively equal to C and D;
A is added to B and C to D; therefore
The sum $A + B$, is equal to the sum $C + D$.

Although the premises of these syllogisms set forth a complexity of relations, the consequents, "shortest possible," "continued parallelism," and "equality of the sums," do not follow the fact that a first thing is related to a third through a second. Notwithstanding the complex antecedents, the sequences are as immediate as that from substance to space, or from event to cause.

Mediate relational syllogisms — or, rather, syllogisms of mediate relativity, always argue that a first thing is related to a third, because it is related to a second which is related to the third. Thus we say,

The line A is parallel to B;
And B is parallel to C; therefore
A is parallel to C.

Things are mediately connected by means of spatial and temporal relations, and also as having quantity and number, as being causes or effects, and as being similar and diverse, identical and different. Hence we reason according to such laws as these:

A contains B; B contains C; therefore A contains C.
A excludes B; B contains C; therefore A excludes C.
A is before B; B is before C; therefore A is before C.
A is contemporaneous with B; B with C; therefore A is contemporaneous with C.
A is greater than B; B is greater than C; therefore A is greater than C.
A is equal to B; B is equal to C; therefore A is equal to C.
A is equal to B; B is less than C; therefore A is less than C.
A is part of B; B is part of C; therefore A is part of C.
A is like B; B is like C; therefore A is like C.
A is like B; B is unlike C; therefore A is unlike C.
A is the same as B; B is the same as C; therefore A is the same as C.
A is the same as B; B is other than C; therefore A is other than C.
A is part of B; B is part of C; therefore A is part of C.
A is like B; B is C; therefore A is like C.

Inferences following such laws as the foregoing are "mediate," because they assert that A is related to C through B. But they are not mediate in the sense that a first thing is antecedent to a third, because it is antecedent to a second, which is antecedent to the third. They do not set forth any second thing which is both consequent and antecedent, but only a first thing (in which two relations combine to form an antecedent) and a second thing (in which a third relation is inferred). In this light they are immediate inferences.

Relational inference, and orthological reasoning in general, need little logical direction. Every argumentative step must be made carefully in accordance with its proper law; that is all. The construction of equations in Algebra and of diagrams in Geometry sometimes require an ingenuity which only nature and practice can supply; but the demonstration which follows calls simply for a clear intelligence. The rules and hints of logic relate chiefly to those syllogizings which pertain to the workings of the material and of the moral universe and to the practical business of life.

4. The homologic syllogism is the explicit statement of any inference based on the homologic principle; for all such inference is double-grounded.

The primary form of it is the paradigmatic — the argument from example — in which one individual sequence is inferred directly from another.

> This powder is poison;
> That powder is exactly like this; therefore
> It also is poison.
>
> In that circle the ratio of diameter to circumference is 3.1416;
> This circle is precisely like that one; therefore
> Its diameter is to its circumference as 1 to 3.1416.

The reasoning thus expressed uses analysis and abstraction, but not generalization. Yet when we dwell on the inference the abstraction runs into generalization; so that argument from example commonly takes the form of reasoning through a generalization. This, however, is not always the case; and paradigmatization should be recognized as the essential type of all homologic inference.

Next, there is the principiative syllogism, in which we infer the general law from the individual sequence.

> This arsenic powder is poison;
> But all arsenic powder is like this in its composition; therefore
> All arsenic powder is poison.

> This circle, by reason of its formation has a fixed ratio between diameter and circumference;
> All circles are formed like this; therefore
> All have that ratio.

> John, Thomas, Peter, *et al.*, die by reason of their constitution;
> All men are constituted like John, Thomas, Peter, *et al.*; therefore
> All are mortal; or (more abstractly) man is mortal.

Sometimes the principiative syllogism is called the inductive; but induction is only the most important species of principiation (Chap. XVI.).

Finally, the most advanced form of homologic sequence gives the applicative, or, as it might also be named, the singularizing, or the individualizing, syllogism. This infers an individual truth from a general principle. It is easily constructed. The major premise asserts a general sequence; the minor ascribes to some individual subject the character of the antecedent of the sequence; the conclusion declares that the consequent follows the individual subject individually.

> Man is mortal;
> Julius Cæsar was a man; therefore
> He was mortal.

This inference presents no practical difficulty; but we bespeak for it careful analysis. For the applicative syllogism, instead of being distinguished from the syllogism proper, has been taken as the type and example of it.

Three things are noticeable in the conclusion, "Cæsar was mortal." First, our thought is changed from the general to the individual, or singular; secondly, our conviction is changed from the hypothetical to the actualistic; and thirdly, a new subject is combined with the predicate of the major premise, "Cæsar" taking the place of the original subject "man." The first two of these changes are justified by the principle that what is true in the general (hypothetically) is true in the particular (actualistically)—which follows from the homologic law as combined with the translative principle; and the new subject is warranted by the principle that what belongs necessarily to any (substantal) predicate belongs also to any subject in which that predicate may inhere. Cæsar being a man, anything belonging to a man, as such, belongs to him. The operation of this simple orthologic principle is scarcely observable, but must be allowed so as to bring "Cæsar" under the homologic reasoning. It does not assume that Cæsar *must* be a man, but only the fact that he is a man.

Turning now to the syllogism,

>Metals are fusible;
>Gold is a metal; therefore
>It is fusible,

we find that the conclusion (*a*) makes no change from the general to the individual, nor (*b*) any from the hypothetical to the actualistic, and (*c*) that the premise "gold is a metal" does not contribute to the conclusion by asserting a fact, but by asserting an hypothetical sequence; the conclusion also being an hypothetical sequence. For any predication that is general (and not merely a collective assertion) sets forth an hypothetical sequence. Clearly this catenate reasoning differs, both in its origin and in its effect, from that inference which applies general truths to existing individuals.

The catenate process may be so conceived as to have a superficial similarity to the applicative inference; it may even be called the application of a general truth to a general subject. Nevertheless, as a mode of inference, it differs radically from that application which is individualizing and actualistic; and which is pre-eminently applicative. It does not follow the law that what is true in the general is true in the individual and actual, but the law that the consequent of a consequent must be a consequent of the antecedent also. It is not the application of principles to realities so as to produce actualistic conviction, but the combination of one general principle with another so as to produce a third.

5. *The inference of the actual from the hypothetical*, which is a factor in applicative syllogizing, *often takes place independently, and gives rise to a syllogism of its own.* This inference may be styled translative, because it transfers the action of the mind from hypothetical to actualistic conviction; but the formal expression of it is known as "the hypothetical syllogism." In this the major premise asserts a sequence hypothetically; the minor either asserts the antecedent of that sequence as actual, or denies the consequent; then the conclusion either asserts the consequent actualistically, or denies the antecedent. The different modes of this syllogism have been already discussed (Chap. XVIII.).

This hypothetical syllogism (along with the applicative) is distinguished from the syllogism proper by reason of the peculiar translative law on which it rests; and it is contrasted with all other syllogisms whatever in that it involves no modification of our conceptions, but only changes the kind of conviction, with which our thought is accompanied.

6. Syllogisms proper arise when two illative propositions are combined so as to produce a third, of which the subject, or antecedent, is taken from one of the original propositions, and the predicate, or consequent, from the other. Moreover, while illative propositions are either singular or general, and may be combined either in the singular or in the general, the syllogisms discussed in logic are those of three *general* propositions. Sometimes, especially in mathematical demonstration, we derive one singular sequence from the combination of two others; yet, even then, when we dwell on a demonstration for the purpose of understanding and testing it, the argument puts on the form of generality, and is expressed in the general.

The syllogism,

> All metals are fusible;
> Gold is a metal; therefore
> Gold is fusible,

is a regular Aristotelian syllogism. So also is this,

> All well-principled persons are trustworthy;
> Some slaves are well-principled; therefore
> Some slaves are trustworthy.

The first of these syllogisms consists of three general necessary sequences. In the second, the first proposition sets forth a general necessary sequence; the remaining two express the general contingent sequences that a slave may be well-principled, and that he may be trustworthy. The law governing these two syllogisms is that *what involves a consequent involves every consequent of that consequent.* This expresses their radical nature. On the other hand, the syllogism,

> All conquerors have strong wills;
> Napoleon was a conqueror; therefore
> He had a strong will,

and the syllogism,

> No conqueror is scrupulous;
> Napoleon was a conqueror; therefore
> He was not scrupulous,

do not follow the law of the consequent-consequent, but are essentially homologic and applicative. The difference of syllogisms proper from these syllogisms appears in connection with the second premise. When we say,

> Metals are fusible;
> Gold is a metal; therefore
> It is fusible,

the second premise does not assert gold to be actually existent; nor does it speak of this or that gold; but it asserts, hypothetically, that if, or whenever, or wherever, there is gold, it is a metal — that the nature "gold" involves the nature "metal." Then, combining this sequence with that of the first premise, we obtain, not an individualized truth, but another general sequence, "gold is fusible." This result may be used in an applicative syllogism concerning this or that gold; but it is quite different from the conclusion of such a syllogism.

The applicative syllogism derives its life and force from the homologic principle; the syllogism proper does not. General illative propositions do, indeed, presuppose principiative inference, as the origin of their generalization; and, in all abstract argumentation, we assume that we can reason in the general, or that general premises will justify a general conclusion. All this rests on the homologic principle. Yet *the vital force of the Aristotelian syllogism is not homological.* We reason in the general — that is, with general antecedents and consequents — just in the same way that we reason in the individual, or with individual antecedents and consequents. The premise, "metals are fusible," uses the antecedent "metal" and the consequent "fusible," and draws its life from their relation as antecedent and consequent; the second premise has "gold" for antecedent, and "metal" for consequent; and, just after the same fashion, the syllogism, "metals are fusible; gold is a metal; therefore gold is fusible," follows the law that the antecedent of a consequent is antecedent also to the consequent of that consequent.

This law is a self-evident corollary, or accompaniment, of the general law of Antecedent and Consequent; it may be briefly named the law of *the consequent-consequent.* But it is not the only principal law of syllogizing proper. There are three others, each of which assumes the law of the consequent-consequent, and is logically dependent on it; and two of which, at least, have an independent operation. These principles may be named the law of "the *separating-consequents,*" the law of "the *common-antecedent,*" and the law of "*syllogistic-reciprocation.*" All three originate from conversional additions to the law of the consequent-consequent; but they can operate independently, because, after any general mode of inference has been discovered, it may be used independently of its origin.

7. The law of the separating-consequents is that if *two antecedents have contradictory consequents, one of the antecedents may be denied of the other, provided that the premise which is to give the consequent of the conclusion can be converted by "the denied-consequent."* This mode of conversion is necessary in order that the antecedent of the converted premise may be denied in the conclusion; "the denied-consequent" being the only kind of conversion which results in denial. Moreover, as only apodeictic propositions can be converted in this way (Chap. XXI.), *the premise to be converted must be apodeictic.* For example, we say,

> No material thing is a free agent;
> Every spirit is a free agent; therefore
> No spirit is a material thing.

This conclusion, being a negative predication, has for its true consequent "not a material thing"; it means "a spirit is necessarily not a material thing"; and this consequent is originally reached by converting the first premise and then combining the converse with the second premise, according to the law of the consequent-consequent. Thus:

By "the denied-consequent," the premise

> No material thing is a free agent,

yields,

> No free agent is a material thing:

then we have the syllogism (of the consequent-consequent),

> No free agent is a material thing;
> Every spirit is a free agent; therefore
> No spirit is a material thing.

This explains the conclusion obtainable by "the separating-consequents."

When both premises are apodeictic, either may be converted by "the denied-consequent"; therefore the antecedent of either may be denied of the antecedent of the other. Thus, converting

> Every spirit is a free agent,

we have

> What is not a free agent is not a spirit;

then, combining this with the other premise of the original syllogism, we have, according to the consequent-consequent,

> What is not a free agent is not a spirit;
> No material thing is a free agent; therefore
> No material thing is a spirit.

In this syllogism the second premise has, for consequent, "not a free-agent"; and this is the antecedent of the other premise.

If, however, either premise be contingent, only the antecedent of the apodeictic premise can be denied of the other antecedent. We can say,

> No vices are praiseworthy;
> Some habits are praiseworthy; therefore
> Some habits are not vices.

But we cannot say,

> No vices are some habits,

or, rather, it would be nugatory and useless to do so. For this conclusion is reached by converting the particular, or contingent, premise by "the denied-consequent"; and it partakes of the worthless character of that conversion (Chap. XXI.). The syllogism producing it would be,

> Some things not praiseworthy are not some habits;
> No vices are praiseworthy; therefore
> No vices are some habits.

The law of the separating-consequents is so named, because the mutual contradiction of the consequents necessitates the conclusion that one of the antecedents is excluded, either absolutely or contingently, from existing in the same subject with the other.

8. The law of the common-antecedent is, that *if two consequents have the same antecedent, either consequent may be asserted contingently of the other.* The operation of this rule requires that one premise only be converted by "the asserted-consequent." Any sequence may be converted in this way; therefore the common-antecedent is a less restricted principle of syllogizing than "the separating-consequents"; which requires the conversion of an apodeictic proposition.

Moreover, when we syllogize according to the common-antecedent, the premise converted does not furnish the consequent, but the antecedent, or subject, of the conclusion. If it be true that

> Some homicides are laudable; and
> All homicides are cruel,

then we can say that

> Some cruel things are laudable.

This conclusion is the "pure," or dogmatic, expression of the half-guarded contingency, "a cruel thing may be laudable." It is obtained, according to the law of the consequent-consequent, after the conversion of the second premise; as follows:

> Some homicides are laudable;
> Some cruel things are all the homicides; therefore
> Some cruel things are laudable.

A similar half-guarded contingency follows from the original premises, if we convert the first, and say,

> Some laudable things are homicides;
> All homicides are cruel; therefore
> Some laudable things are — or a laudable thing may be — cruel.

In converting a negative premise in any syllogism of the common-antecedent, we must remember that the consequent of a negative sequence is not expressed by the predicate term alone, but *by that term along with the negative particle*. Hence, according to the law of the "common-antecedent," the premises,

> No moral precept is a material thing; and
> All moral precepts are useful,

yield both the following conclusions:

> Some useful things are not material; and
> Some non-material things are useful.

The first of these conclusions evidently follows, according to the consequent-consequent, after the conversion of the affirmative premise by the "asserted-consequent"; the second follows, just in the same way, after converting the negative premise by the "asserted-consequent." For inspection shows that the consequent of the negative premise is "not material"; and, converting with this consequent, we say,

> Some things not material are moral precepts;
> All moral precepts are useful; therefore
> Some things not material are useful;

which is a syllogism of the consequent-consequent.

9. The third subordinate, or, more simply, the fourth law of catenate inference, is that of syllogistic-reciprocation. It

is essentially double; one part of it pertains to affirmative, or conjunctive, reciprocation, and the other to negative, or disjunctive, reciprocation.

The law of affirmative reciprocation is that "*if the consequent of a first sequence be antecedent in a second sequence, then the consequent of that second sequence may be made antecedent of contingency to the antecedent of the first sequence.*" Here we call that premise the first whose consequent-term is the antecedent-term of the other premise; and we call the other premise the second. This use of terms will be maintained throughout our discussion of the laws of syllogistic-reciprocation.

Both these laws of reciprocation may be explained as conditioned on the principle of the consequent-consequent; but they differ from the subordinate laws already considered, in that their explanation involves the conversion of both premises. Take, for example, the affirmative sequences,

> Some virtuous men are necessitarians; and
> All necessitarians are speculators.

Converted, both by the asserted-consequent, — the latter with the retained necessitant, — and reversing the order of the premises, we have, by the consequent-consequent,

> Some speculators are all the necessitarians;
> Some necessitarians are virtuous men; therefore
> Some speculators are virtuous.

This may illustrate the affirmative law.

The negative law is not so simple. It is that "*if, in two consecutive sequences, opposite to each other in quality, the predicate of the first be the subject of the second, a new negative sequence may be formed with the predicate of the second sequence for subject and with the subject of the first sequence for predicate; provided, however, the first sequence* (whether affirmative or negative) *be apodeictic, and provided also that the negative sequence be apodeictic* (whether it be first or second)." More briefly, the first (which is, in this case, the "major") premise is always apodeictic; the second (which is, in this case, the "minor") premise must be apodeictic in case it is the negative one — otherwise it may be contingent.

The negative premise, whether first or second, *must be apodeictic*, because a negative conclusion can be obtained only through converting that premise by the denied-consequent; which principle applies only to apodeictic propositions. *The first premise must be apodeictic*, because that premise, after conversion, is to furnish the predicate of the negative conclusion. Were it contingent, the conclusion could not have an absolute (or distributed) predicate; and would, therefore, be useless; like the "simple" converse of a particular, or contingent, negative.

To illustrate: in the following syllogism, the first (or major) premise is the negative one, and therefore a contingent minor is admissible.

> No moral motivity is an animal impulse;
> Some animal impulses are principles of action; therefore
> Some principles of action are not moral motivities.

Here "principle of action" becomes, by the asserted-consequent, antecedent of contingency to "animal impulse"; then, by the denied-consequent, "animal impulse" becomes antecedent of impossibility to "moral motivity"; and so (changing the order of premises) we reason, by the consequent-consequent, thus:

> Some principles of action are animal impulses;
> No animal impulse is a moral motivity; therefore
> Some principles of action are not moral motivities.

In the following syllogism, the second (or minor) premise is negative, and must, therefore, be apodeictic:

> All ruminants have four stomachs;
> No four-stomached animal is carnivorous; therefore
> No carnivores are ruminants.

Were the second premise here a particular negative, it could not be converted by the denied-consequent, so as to assert "not four-stomached," the contradictory of the consequent of the major premise; without which assertion there could be no negative conclusion. But the consequent of the major (four-stomached) being thus denied, its necessitant (ruminant) can be denied in the conclusion.

This last syllogism (like all reciprocative syllogisms) assumes the consequent-consequent form when we convert both premises, and reverse their order; thus,

> No carnivores are four-stomached;
> Some four-stomached are all the ruminants; therefore
> No carnivores are ruminants.

It also brings before us another instance in which conversion "per differentiam," or with "the retained necessitant," is necessary to a valid conclusion.

10. While both modes of reciprocative syllogizing may be accounted for as an operating of the law of the consequent-consequent after two conversions, it must be added that probably such a process is never carried out in our ordinary and spontaneous thinkings. Therefore, also, it is yet more unlikely that any mind ever uses either law of syllogistic reciprocation — but especially the negative one — independently of its origin, or mode of formation. Aristotle seems to have been right in recognizing only three normal forms of syllogizing. This catenate reciprocation is at best an occasional and accidental — not a spontaneous and natural — mode of inference. It has the appearance of originating in an effort, which cannot be directly carried out, to syllogize according to the consequent-consequent.

This led Sir Wm. Hamilton to say that the fourth "figure" is a distorted form of the first. But we judge that reciprocative arguments are more frequently completed by the methods of the "separating consequents," and the "common-antecedent," than in any other way. Positive syllogisms may be completed by the latter method, if we only convert the first (or major) premise; and negative by the former, if we only convert the second (or minor) premise. Take, for example,

> All greyhounds are dogs;
> All dogs are quadrupeds; therefore
> Some quadrupeds are greyhounds.

Converting the major, we have,

> Some dogs are greyhounds;
> All dogs are quadrupeds; therefore
> Some quadrupeds are greyhounds,

which is according to the "common-antecedent."

Take also the negative syllogism,

> All ruminants are four-stomached;
> No four-stomached animals are carnivores; therefore
> No carnivores are ruminants.

Converting the second (or minor) premise, we have,

> All ruminants are four-stomached;
> No carnivores have four stomachs; therefore
> No carnivores are ruminants,

which is according to the "separating-consequents."

Such, then, are the four fundamental modes of catenate inference; in one or other of which every act of syllogizing takes place. We shall soon consider according to what laws conclusions are sometimes affirmative, and at other times negative; also sometimes universal (or apodeictic), and at other times particular (or contingent). Let us now complete our general survey of the syllogism by defining its essential parts, and their properties.

11. First, *a syllogism consists of three illative propositions, and only three.* This follows from the very nature and definition of catenate inference; and is manifest in connection with each of the four laws of syllogizing. The two propositions, which, in combination, constitute the syllogistic antecedent, are called "the premises"; the third proposition, which sets forth the consequent of that antecedent, is "the conclusion." One of the premises furnishes the subject of the conclusion, and is called "the minor premise"; the other furnishes the predicate of the conclusion, and is called "the major premise."

In the next place, *every syllogism contains three "terms," or "extremes," and only three.* Verbally, these terms are the general names, or nouns, or nominal expressions, used as subjects or predicates in the propositions; mentally, they are general notions, or conceptions. They are called "terms," or "extremes," because a proposition may be symbolized by a line with the subject at one end and the predicate at the other.

Only three terms are admissible, and three are requisite, according to the essential law of catenate inference — the law of the consequent-consequent. According to this law, the consequent of the minor premise (either at first, or after such conversion as may be necessary) is also antecedent of the major; and then the antecedent of the minor and the consequent of the major form the conclusion.

Hence, also, one term is always common to both premises. This is known as "the middle term"; because, in the natural order of "the consequent-consequent," it comes between the other two terms. Of these one is common to the minor premise and conclusion, and is called "the minor term"; the other is common to the major premise and conclusion, and is called "the major term." The terms which become subject and predicate of the conclusion are designated "minor" and "major"; because, in constructing the most common syllogism — the affirmative syllogism of the consequent-consequent — we generally conceive of the major term as having wider "extension," or application, than the minor. In saying,

> Men have rights;
> Slaves are men; therefore
> Slaves have rights,

the term "slave," in the premises and in the conclusion, has less extension than "have rights." This mode of conception is by no means necessary, and does not belong to every syllogism; therefore the designations "major" and "minor" are somewhat arbitrary.

12. *In analyzing a syllogism one should begin with the conclusion,* or the proposition to be proved. The order of enunciation does not reveal which premise is major nor which is minor; either may be enunciated first. The conclusion also may either precede the premises, or follow them, or come between them. But, in every case, the predicate of the conclusion is the major term, and the subject of the conclusion the minor term; then that proposition *which contains the predicate of the conclusion* (together with the middle term) is the major premise; and that *which contains the subject of the conclusion* (with the middle term) is the minor premise.

13. Thirdly; the "figure" of a syllogism is its character with reference to the place of the middle term in each premise.

That term may be subject of the major, and predicate of the minor; then the syllogism is of the *first figure:* — it may be predicate of both premises; then the syllogism is of the *second figure:* — it may be subject of both premises; then the syllogism is of the *third figure:* — or it may be predicate of the major, and subject of the minor; and then the syllogism is of the *fourth figure*. Employing the letters P, S, and M for the major, minor, and middle terms, and placing the major premise first, the figures are as follows:

	Fig. I.	Fig. II.	Fig. III.	Fig. IV.
Major premise	$M-P$	$P-M$	$M-P$	$P-M$
Minor premise	$S-M$	$S-M$	$M-S$	$M-S$
Conclusion	$S-P$	$S-P$	$S-P$	$S-P$

All syllogisms of the consequent-consequent necessarily assume the first figure; and are known as syllogisms of that figure. In like manner syllogisms of the separating-consequents necessarily assume the second figure; those of the common-antecedent, the third figure; and those of syllogistic-reciprocation, the fourth.

The question whether major or minor premise should be enunciated first, has been greatly discussed. It should be answered by saying that no absolute rule can be justified. In the order of investigation and inferential discovery, the minor premise comes first; for that premise contains the subject of enquiry and assertion. But in argument and controversy, the major premise is the more prominent. Such, at least, is the case with syllogisms in the first three figures; which are the only figures in which we reason spontaneously; and which alone were recognized by Aristotle.

The fourth figure belongs to a kind of accidental syllogizing, in which we set out, or attempt, to use the first figure, and then form a conclusion by the aid of conversion. The order of its enunciation is subject to the same influences as that of the first figure; and is the same as it would be, if we could complete our reasoning without conversion. Hence, in the first three figures, the order of discovery places the minor premise first; and the order of argument, the major first; while in the fourth figure this rule is reversed. The fourth figure uses for its minor premise what would be the major premise in the first; and for its major, what would be the minor.

Moreover, in formal demonstration, we know what we have to prove, and may mention it first, if we like. Therefore, in every figure, the conclusion may be stated either before or after the premises. Aristotle and the Greek logicians did not confine themselves to one order of enunciation. They often placed the minor premise first; and sometimes, the conclusion. The scholastics and the moderns have favored what is called "the synthetic order"; in which the major premise precedes the minor premise and the conclusion. Although this order (except in the fourth figure) does not place the middle term in the middle of the process of thought, and is therefore secondary and artificial, it presents arguments with clearness and force; wisdom also suggests that it be adopted for the sake of uniformity; and to avoid confusion.

14. Finally, the "mood" of a syllogism is *its character with reference to the quality and quantity of its three propositions*. The Greek logicians called this the syzygy (συζυγία)—the combination, or "conjugation"—of a syllogism. Of course each of the three propositions may be either affirmative or negative; and at the same time, also, each may be either universal or particular. In symbolic language, each may be either *A*, *E*, *I*, or *O*. The mood of a syllogism is stated by using these symbols to indicate the character of the three propositions. *AAA* is the mood of a syllogism all whose propositions are universal affirmatives. *EAE* is a mood in which the major premise is an universal negative, the minor an universal affirmative, and the conclusion an universal negative. *IAI* is a mood in which the major is a particular affirmative, the minor an universal affirmative, and the conclusion a particular affirmative. One of the principal investigations of logic determines what moods, and how many, are valid, in each of the four figures.

The statement that the mood of a syllogism lies in the quality and quantity of its propositions, applies only to syllogisms composed of pure, or dogmatic, propositions; and which, therefore, are styled pure, or dogmatic, syllogisms. But the quantity of a dogmatic statement is only the superficial expression of its modality; universal quantity indicates necessity, particular quantity, contingency. Therefore, to define *mood* by its relation to internal and mental assertion (ὁ ἐν τῇ ψυχῇ λόγος), we must say that it *is the character of a syllogism with reference to the quality and modality of its propositions.*

This conception was that which Aristotle entertained; and which he carried out with laborious fidelity. It has the advantage of being applicable to all syllogisms whatever, both to those composed exclusively of dogmatic assertions, and to those constituted wholly, or in part, of modal predications. Moreover, as it is not limited to any specific form of statement, but pertains to the essential nature of catenate inference, it will prepare us to admit, and to understand, certain delicate syllogistic conclusions which cannot be expressed dogmatically.

CHAPTER XXIII.

SYLLOGISTIC MOODS.

1. If the mood is valid, the syllogism is valid. 2. Any sequence whatever can be converted by the asserted-consequent. 3. Definitional conversion. 4. Only apodeictic propositions admit the denied-consequent or the retained-necessitant. 5. In syllogisms a negative proposition must sometimes be taken as affirmative. 6. Under the consequent-consequent, (*a*) if either premised sequence be contingent, the conclusion must be contingent; (*b*) if the second sequence (major premise) be contingent, the conclusion will be unguarded. 7. An exception to this last rule. 8. The valid moods of the first figure: *AAA, EAE, AII, EIO*; and *IAI, OAO, III, OIO*. 9. Of the second figure: *AEE, EAE, AOO, EIO*; and *AAI, AII, AII, III*. 10. Of the third figure: *AAI, AII, EAO, EIO, IAI, OAO*; and *III, OIO*. 11. Affirmative moods of the fourth figure: *AAI, IAI*; and *AII, III*. 12. Negative moods of the fourth figure: *AEE, EAO, EIO*. 13. Contingent premises are commonly guarded. *In all the mood-formulæ they are assumed to be guarded.* For any syllogism with an unguarded contingent premise must have an unguarded conclusion, no matter what be its figure or mood. 14. The negative moods of the fourth figure appeal to the separating-consequents; and its positive moods, to the common-antecedent. 15. The consequent-consequent and the separating-consequents are the dominant laws of catenate syllogizing.

1. WE ascertain the specific forms of correct syllogizing by determining what moods are valid in the different figures. For to say that a certain mood is valid in any figure, is to say that two premises of given quality and modality will produce a correct conclusion of given quality and modality. To do this with a thorough intelligence, one's thoughts should not be confined to those dogmatic propositions which set forth the reciprocal inclusions and exclusions of logical classes, or to syllogisms constructed from such assertions; the internal and modal syllogism should be the subject of our investigations. Pure predications excellently set forth the most prominent modes of logical sequence, yet they are secondary forms of thinking,

and should ever be accompanied by mental interpretation. Every syllogism, however immediately conceived, should be regarded as constituted of three general sequences; and the laws of syllogizing should be formulated with reference to this essential doctrine.

2. Therefore, in attempting this formulation, some principles respecting illative statements must be borne in mind.

For example, *in converting propositions, we must deal with them as composed of antecedent and consequent*, rather than as composed of subject, copula, and predicate; and we must apply the laws of "the asserted-consequent," the "denied-consequent," and the "retained-necessitant."

Any sequence whatever may be converted by the asserted-consequent; and will then have an affirmative contingent proposition for its converse. "A horse is (necessarily) a quadruped," yields "a quadruped may be a horse"; "a horse may be wild," yields "a wild animal may be a horse"; "a horse has no horns," yields "an animal without horns may be a horse"; "a horse may not be sound," yields "an animal not sound may be a horse." The converse of an affirmative sequence, is an affirmative sequence with a positive subject; the converse of a negative sequence, is an affirmative sequence with a negative subject.

It may be said that this conversion of negatives does not proceed directly, but is conditioned on a substantialization of the consequent, whereby the convertend really becomes an affirmative. "A horse may not be sound," becomes first "a horse may be an animal not sound"; and is then converted in the same way as an affirmative. This is true; but this does not show that negatives are not convertible by the asserted-consequent; it shows how they are converted by that method.

3. One exception must be noted to the rule that the asserted-consequent produces a contingent converse. *When the antecedent is an exact necessitant of the consequent, an apodeictic converse may be asserted.* If the elephant is the largest of quadrupeds, the largest of quadrupeds is the elephant. Definitions and certain mathematical inferences may be dealt with in this way (Chap. XX.).

4. Conversion by the denied-consequent occurs most fre-

quently with the universal negative; but *may be used with any apodeictic proposition, whether affirmative or negative.* "No men are perfect," which means "a man cannot be — or is necessarily not — a perfect being," yields "no perfect beings are men," or "a perfect being is necessarily not a man." Here "perfect," the antecedent of the converse, is contradictory of "not perfect," the consequent of the convertend. In the same way, "all men are fallible," yields "no infallible beings are men," "infallible" being the contradictory of "fallible." This mode of conversion always produces an apodeictic negative; it asserts that the antecedent cannot exist when its necessary consequent is denied (Chap. XV.).

Contingent propositions reject conversion by the denied-consequent (Chap. XXI.).

The retained-necessitant is a specific mode which the asserted-consequent may assume in the case of apodeictic propositions. It produces an affirmative contingent sequence, but has the peculiarity that the consequent of the converse retains the same absoluteness of application which it had as the antecedent of the convertend. Ordinarily, "all men are mortals," yields "some mortals are men," that is, "a man may be a mortal"; instead of this, with the retained-necessitant, we say, "Some mortals are all the men," or "only mortals are men," or "mortals include all the men," or "only a mortal can be a man." This conversion being founded on the fact that "mortal" is a necessary, not an accidental, ascript of man, it may be styled conversion "*per differentiam*"; provided the word "*differentia*" be taken to signify any necessary characteristic.

Negative necessity is sometimes converted with the retained-necessitant; though not so frequently as positive necessity. The result of such conversion, in conformity with the general operation of the asserted-consequent, is an affirmative contingency with a negative antecedent. "No men are perfect," yields "only imperfect beings are men," or "only an imperfect being can be a man," or "an imperfect being, differentially, may be a man."

That retained *absoluteness of conception*, which sometimes appears in the predicate after the conversion of necessities by the asserted-consequent, *is always retained after conversion by the denied-consequent.* This latter principle allows no option between two styles, or degrees, of conversion. The antecedent of the original proposition must be absolutely rejected; else there would be no usable converse. After "no man is perfect" is converted into "no perfect being is a man," the predicate "man" retains that absolute force with which, in the convertend, it renders "perfect" impossible. The impossibilitant, the necessitant of negation, is retained.

5. With respect to *the combination of propositions as affirmative and negative,* the following rule provides for simplicity of statement; viz., a negative sequence is to be classed with affirmatives, whenever it must assume a positive form before being connected with the other sequences of a syllogism. For example, should we say,

> Wood is not metallic;
> What is not metallic cannot be used as coin; therefore
> Wood cannot be used as coin,

the minor premise, "wood is not metallic," is negative, yet, before combining it with the major, we give it an affirmative form by mentally substantializing its consequent, and saying, "Wood is a thing not metallic." This change is necessary in order that the consequent of the minor may become the antecedent of the major. In such cases we say that the minor premise must be affirmative; though this is not literally true. The exact statement is that it must be affirmative, or, if it is negative, that it must be given an affirmative form; so that its consequent may agree with the antecedent of the major. For an antecedent conception, even though essentially negative, always assumes a positive form.

The above rule, respecting negative sequences, qualifies and interprets the common teaching that affirmative conclusions require both premises to be affirmative; and that negative conclusions require one premise to be affirmative and the other negative. Should we say,

> What is not truly valuable is not sought by the wise;
> The applause of the world is not truly valuable; therefore
> It is not sought by the wise,

we have a negative consequent from two negative premises. And should we say,

> What is not compounded is an element;
> Hydrogen is not compounded; therefore
> It is an element,

we have an affirmative conclusion with one of the premises negative. But in each of these syllogisms the minor premise must be classed with affirmatives.

6. With respect to *the combination of premises as necessary and contingent*, the following principles should be remembered. They apply directly only to syllogisms of the consequent-consequent, but indirectly to all syllogisms.

First, when both premises are apodeictic, the conclusion is apodeictic; but if either be contingent, the conclusion must be contingent. Moreover, as certainty may be indicated by unity, contingency by the ratio of the chances, and the likelihood of a compound sequence by the product of the probabilities of its parts (Chap. XIX.), a syllogism with one contingent premise has a conclusion of the same degree of contingency with that premise; while, if both premises of a syllogism be contingent, the conclusion is weaker than either premise. In the argument, "Robbery may lead to murder; and murder, to hanging (or death by electricity); therefore robbery may lead to hanging," the contingency of the conclusion would be equal to the product of the fractions representing the separate probabilities of the premises; if those fractions could be ascertained.

Secondly, the style of the contingency of a conclusion as guarded or unguarded — which is a matter of more consequence than the degree of the contingency — may be determined by a rule in which the minor premise, according to the natural order, is conceived of as the first, and the major as the second. This rule, of course, applies only to cases in which one of the premises, at least, is contingent. It is that *if the second sequence be apodeictic, the conclusion will have the same style of contingency as the first, but if the second sequence be contingent, the conclusion will be an unguarded contingency.* In other words, if the major premise be apodeictic, the conclusion will have the same contingency, whether guarded or unguarded, as the minor premise, but if the major be contingent, the conclusion will be an unguarded contingency, no matter what may be the character of the minor. For instance, the syllogism,

>   A carnivore may be a lion (minor);
>   A lion is a quadruped (major); therefore
>   A carnivore may be a quadruped,

has a conclusion guarded against impossibility; because the minor is so guarded, and the major is apodeictic. But should we say,

>   A lion is a carnivore (minor);
>   A carnivore may be a bird (major); therefore
>   A lion may be a bird,

the conclusion, though a correct conjectural judgment, would be unguarded, the major premise not being apodeictic.

The reason on account of which, in order to a guarded conclusion, the consequent of the prior sequence must be an antecedent of necessity in the second sequence, is that otherwise the antecedent given in the prior sequence might be found to be wholly excluded from participation in the second sequence. Let A be necessarily, or contingently, B; and B contingently C. Both these things may be true, while yet in every case in which A is B, A is not, and cannot be, a B which is a C. Therefore a guarded contingency follows only when the minor premise sets forth a guarded contingency, and the major is apodeictic.

7. An exception to this requirement of an apodeictic major occurs *whenever the minor premise has an absolute, or necessitant, predicate.* This happens not only in exact, or reciprocative, necessitations, but also in other cases, such as exist after conversion with the retained-necessitant. For example, should we say,

>   Some books are novels (minor);
>   Some novels are morally injurious (major); therefore
>   Some books are morally injurious,

the conclusion would be guarded, because all novels are books. An abstract statement of this argument would be,

> A is differentially B;
> B is contingently C; therefore
> A is contingently C.

Here the contingency of the conclusion must be guarded if that of the major is; because, every B being an A, A must certainly participate in the contingent relations of B.

8. We are now prepared to say what moods — or combinations of propositions — are valid in the first figure; or, more explicitly, in syllogisms of the consequent-consequent. We shall speak first of quality; and then of modality.

As regards quality, (*a*) *the conclusion of a syllogism in the first figure must agree with the major premise;* for it always asserts the consequent of the major, whether affirmative or negative, as following the antecedent of the minor. (*b*) *The minor premise must always be affirmative.* But this means only that if the minor happen to be negative, it must be given a positive form. (*c*) Finally, since *the major premise* may set forth any general sequence, that premise *may be either affirmative or negative.*

With respect to modality, the principle of the consequent-consequent allows any combination of premises as apodeictic and contingent, with the following restrictions. (*d*) *If both premises be apodeictic, the conclusion is apodeictic.* (*e*) *If either or both be contingent, the conclusion is contingent.* But (*f*) *in order to infer a guarded contingency, the major premise must be apodeictic.*

In the following syllogism, the major premise being contingent, the conclusion is unguarded:

> One who steals may be caught (minor);
> One who is caught may be punished (major); therefore
> One who steals may be punished.

In this syllogism, were the major premise negative, the conclusion would also be negative; and would assert "may not be punished."

Such reasonings cannot be expressed dogmatically. We cannot say,

> Some who steal are caught;
> Some who are caught are punished; therefore
> Some thieves are punished,

because it may be that none of the caught ones who have stolen are among the caught ones who are punished. Such arguments are declared invalid by those who recognize only pure syllogisms. They are not invalid. They are correct conjectural inferences; and are often used respecting matters of probability.

Admitting them, the major premise may be either A, E, I, or O; while the minor (being affirmative) must be either A or I. Combining major and minor accordingly, and adding the required conclusions, we have the following mood-formulas; in which, according to the common practice, the major premise is indicated first, the minor next, and the conclusion last:

$$AAA,\ EAE,\ AII,\ EIO,\ \text{and}$$
$$IAI,\ OAO,\ III,\ OIO.$$

The first four of these, having apodeictic majors, can produce guarded conclusions, and can be stated dogmatically: the remaining four produce unguarded contingent conclusions; such as that respecting the punishment of the thief.

An unguarded conclusion is insufficient for the refutation of an apodeictic statement. An unguarded O is not the contradictory of A; nor an unguarded I of E. In this sense the arguments producing these conclusions are inconclusive. But this does not justify the rejection of the unguarded moods; it only limits their use to conjectural reasonings.

9. The controlling law of the second figure is that of the separating-consequents. *This finds two antecedents which have contradictory consequents, and then denies one antecedent of the other.* Hence, as to quality, (a) *one premise must be affirmative and the other negative;* and (b) *the conclusion must be negative.* But this last means only that the immediate form of the conclusion must be negative. If the antecedent of the major be essentially a negative conception, the conclusion, as asserting the contradictory of that, is essentially affirmative. In the syllogism,

No unthinking entity is a free agent (major);
All spirits are free agents (minor); therefore
No spirit is an unthinking entity,

the conclusion really signifies

> Every spirit is a thinking entity.

In short, this rule requiring a negative conclusion is similar to that requiring an affirmative minor in the first figure.

With respect to modality, (*c*) *the major premise must be apodeictic.* Were it not so, its antecedent could not be wholly rejected as the predicate of the conclusion; without which rejection there could be no true negation. To say,

> An animal may be a carnivore;
> A horse is not a carnivore; therefore
> A horse may not be an animal,

shows only that a horse may not be some kind of animal — not that it may not be an animal. So,

> An animal may not be a carnivore;
> A lion is a carnivore; therefore
> A lion may not be an animal,

gives the same sort of useless conclusion. Then (*d*) *the minor premise,* as simply furnishing subject and mode of sequence for the conclusion, *may be either apodeictic or contingent.* Finally, (*e*) *the conclusion agrees in modality with the minor.* For the antecedent of the minor supports the contradiction in the conclusion exactly with the force with which it supports its own contradicting consequent.

These rules (*a, b, c, d, e*) require the major to be either *A* or *E*; and allow the minor to be either *A, E, I,* or *O,* provided it differs in quality from the major. Hence, syllogisms of the separating-consequents have only the following four valid moods: *AEE, EAE, AOO, EIO.*

But while this is true, it is not absolutely correct to say that no other moods than these are admissible in the second figure. These are the only negative moods; in addition to these, certain weak affirmative moods, with the middle term predicate in both premises, may be justified by a principle of their own. For *if two antecedents have a common positive consequent, either may be affirmed of the other, though with an unguarded contingency.* The strongest conclusion obtainable in this way from ordinary sequences follows from two apodeictic premises. We may say,

> Horses are animals (major);
> Quadrupeds are animals (minor); therefore
> A quadruped may be a horse;

that is, any quadruped of whose specific character we are ignorant, may be a horse. Notwithstanding the apodeictic premises, this conclusion is an unguarded contingency, because it depends on that conversion of the major, which leads to the following syllogism of the consequent-consequent:

> Some animals are horses;
> All quadrupeds are animals; therefore
> A quadruped may be a horse.

The argument, therefore, is equivalent to one in the first figure with a contingent major; in which, as we have seen, the conclusion is unguarded.

In these syllogisms of the common-consequent, the premises must be affirmative; because nothing could be inferred if the common-consequent were negative. That neither a horse nor a quadruped is a stone, does not warrant even a conjecture that the one is, or is not, the other. But very weak conclusions follow with one premise, or both, contingent. Accordingly we have the following affirmative moods: *AAI, AII, IAI, III.*

Propositions with a common-consequent are easily convertible into propositions with a common-antecedent; and the contingent connection of things is more naturally and fully inferable in connection with a common-antecedent than in connection with a common-consequent. Therefore the affirmative moods of the second figure may be safely neglected, not as incorrect, nor even as abnormal, but as weak and needless.

10. The third figure is governed exclusively by the law of the common-antecedent. Hence the moods of this figure may be determined, if we remember how *the common-antecedent is essentially the consequent-consequent, as operating in connection with a conversion of the minor.* For this premise must be converted by the asserted-consequent, in order that, after conver-

sion, it may have a consequent identical with the antecedent of the major. Taking the syllogism,

>All homicides are cruel ;
>Some homicides are laudable ; therefore
>Some laudable things are cruel ;

and converting the minor, we have,

>All homicides are cruel ;
>Some laudable things are homicides ; therefore
>Some laudable things are cruel ;

which syllogism of the consequent-consequent arises from that conversion by the asserted-consequent; and could not be obtained otherwise.

Now, no negative premise can be converted by the asserted-consequent, unless it be first given an affirmative form. Hence one rule of the third figure is that (*b*) *the minor premise must be affirmative;* by which we mean only that, if that premise happen to be negative, it must be given an affirmative form. In the syllogism,

>All moral precepts are useful ;
>No moral precepts are material ; therefore
>Some things not material are useful,

the minor premise is negative, but must be classed with affirmatives; because the conclusion depends on its affirmative converse, that

>Some non-material things are moral precepts.

After the conversion of the minor, the conclusion adopts and asserts the consequent of the major; hence (*c*) *the conclusion agrees in quality with the major;* while, as in syllogisms of the consequent-consequent, (*a*) *the major may be either affirmative or negative.*

Modality, in the third figure, is determined as follows: First, while (*d*) the premises may be either apodeictic or contingent, (*e*) *the conclusion is always contingent.* Even in the case of both premises being apodeictic, the conversion of the minor by the asserted-consequent renders that premise contingent;

and thus causes a contingent conclusion. Secondly, (*f*) *in order that a guarded contingency may be inferred, either the major premise must be apodeictic*, as in the first figure, *or, should the major be contingent, the minor must be apodeictic.* When the major is apodeictic, its antecedent, after the conversion of the minor, binds the premises together, so as not to allow an unguarded conclusion; and when the minor is apodeictic, its antecedent performs the same part, after the conversion of that premise with the retained-necessitant. The former case does not differ materially from that of the first figure; the latter may be illustrated as follows:

> Some homicides are laudable;
> All homicides are cruel; therefore
> Some cruel things are laudable.

Here the conclusion, as guarded, depends on the differential, converse of the minor — "some cruel things are all the homicides." For, this being granted, it is plain that those cruel things which are "some of the homicides" must be laudable; in other words, that it is not an impossibility, but an absolute possibility, a guarded contingency, that a cruel thing should be laudable. This use of the retained-necessitant is not called for in syllogisms of the consequent-consequent, but often occurs in connection with the conversions of the subordinate figures.

Recapitulating the foregoing rules (in a proper order), we say that, as to quality, (*a*) the major may be either affirmative or negative, (*b*) the minor must be affirmative, and (*c*) the conclusion must agree with the major. As to modality, (*d*) each premise may be either apodeictic or contingent, (*e*) the conclusion must be contingent, and, (*f*) to produce a guarded conclusion, either the major or the minor must be apodeictic.

Combining these rules, we find that the major premise may be either *A*, *E*, *I*, or *O*; the minor may be *A* or *I*; but in case the major is *I* or *O*, the minor must be *A*; and the conclusion must be either *I* or *O*. Accordingly, in the third figure, the valid moods with guarded conclusions are *AAI*, *AII*, *EAO*, *EIO*, *IAI*, *OAO*. But we must recognize also syllogisms with both premises contingent, and whose conclusions, therefore, are unguarded; hence, neglecting the last rule, we form the moods *III* and *OIO*. The syllogism,

> Some men are intelligent;
> Some men are unprincipled; therefore
> Some unprincipled persons may be intelligent,

is in the mood *III*.

11. We pass to the fourth figure, with its two laws of conjunctive, and of disjunctive, reciprocation. The former of these asserts that *if P be the antecedent of M, and M of S, then S is antecedent of contingency to P*. This calls for a syllogism in which the same term, *M*, is both consequent of the major and antecedent of the minor. But an antecedent is always a positive conception; that is, it is either naturally positive or is given a positive form. In this sense, therefore, the consequent of the major must be positive, and (*a*) *the major must be an affirmative proposition*. Notwithstanding this, the major is sometimes essentially negative, as in the following:

> Some virtuous persons are not amiable;
> Persons not amiable have few friends; therefore
> Some persons with few friends are virtuous.

In the same manner, (*b*) *the minor premise must be construed as affirmative;* because its consequent is to be used as antecedent of the conclusion. Yet this premise, also, may be essentially negative, as in the following:

> Some who are respected are hypocrites;
> No hypocrites deserve respect; therefore
> Some who do not deserve respect receive it.

And finally, (*c*) *the conclusion must be affirmative;* because it has for consequent the antecedent of the major. Yet it may be really negative, if that antecedent is a negative conception; as in the following:

> Some persons not virtuous are amiable;
> Amiable persons have many friends; therefore
> Some who have many friends are not virtuous.

With regard to modality, conjunctive reciprocation imposes no restriction on the premises. This mode of syllogizing merely calls for premises which can be converted by the asserted-consequent; by which method all sequences whatever are convertible. Hence (*d*) *both major and minor premise may be either apodeictic or contingent.* But a converse produced by the asserted-consequent is always contingent; and therefore — since nothing but contingency can come from contingency — (*e*) *the conclusion of a conjunctive reciprocation must be contingent.* Indeed, it is always a weak contingency, being the product of two contingencies.

The style of the contingency of the conclusion, however, varies with the character of the premises. If the converse of the major be guarded against impossibility, or necessity of the opposite, (as happens when the major itself is so guarded,) and *if the minor premise be apodeictic,* the conclusion is guarded. Otherwise it is not. For example, should we say either "all pious persons," or,

> Some pious persons, are over-exact;
> All over-exact people are unpleasant company; therefore
> Some persons unpleasant in company are pious,

the conclusion would be guarded. For, the major having been converted by the simple asserted-consequent and the minor differentially, the retained-necessitant of the minor ("over-exact") binds the premises together so as to prevent an unguarded conclusion. The retained-necessitant operates here precisely as in certain moods of the third figure, and, in this case as in that, the law of its operation, expressed abstractly, is that "whatever is contingent (or is in any other way logically related) to any subject, is similarly related to whatever necessarily inheres in that subject." "Pious" being connected with "over-exact" by a guarded contingency, must be similarly related to "unpleasant company," which necessarily inheres in "over-exact." Were the minor premise contingent, this result would fail for the want of a retained-necessitant to bind the premises together. Hence, in affirmative reciprocation, (*f*) *the conclusion is an unguarded contingency, unless the major premise be guarded and the minor be apodeictic.*

Combining the foregoing rules (*a, b, c, d, e, f*), we find that the major premise may be either *A* or *I*; while the minor must be A; and the conclusion I. This allows only two moods, AAI, IAI. But, admitting the unguarded conclusion with a contingent minor, we have two moods more, AII, III. The following is in the mood III:

> Some intelligent beings are men;
> Some men are unprincipled; therefore
> Some unprincipled beings may be intelligent.

12. In disjunctive syllogisms of the fourth figure (a) *the conclusion*, of course, *is negative;* disjunction is negation, or a form of negation. (b) Hence *the premises must be opposite in quality;* because this is always necessary in order to a negative conclusion.

The modality of the premises is affected by the fact that both premises have to be mentally converted; so that the subject of the major may become predicate of the conclusion, and the predicate of the minor, subject of the conclusion. This being so, (c) *the major premise must be apodeictic*, that its subject, as predicate of the conclusion, may have an absolute distributive and exclusive force. Otherwise the negation of the conclusion would be nugatory. For a kindred reason, (d) *in case the major premise be affirmative, the minor must be an apodeictic negative.* Were it a contingent negative, its converse would be nugatory, and without logical force. The following syllogism is inconclusive, because the minor premise is not apodeictic:

> All metals are minerals;
> Some minerals are not poisons; therefore
> Some poisons are not metals.

But (e) *if the major premise be an universal negative, the minor may be either apodeictic or contingent.* For in that case the minor would yield a true contingent converse, and the middle term, if not distributed in the conversa of both premises, would yet be distributed in the converse of the major. Thus,

> No negro is a Hindoo;
> Some Hindoos are black; therefore
> Some blacks are not negroes.

Combining the foregoing rules we find that the major premise may be A or E: if it is A, the minor must be E; but if it is E, the minor may be A or I.

This allows only three pairs of premises, AE, EA, EI. Only the first pair, after conversion, justify an universal conclusion according to the consequent-consequent; in both the other combinations the contingent converse of the minor necessitates a contingent conclusion. We have, therefore, in all, three negative moods, AEE, EAO, EIO.

The first of these may be said to express the specific principle that "*whatever necessitates an entity, cannot inhere in whatever that entity renders impossible.*" Let "lion" render "carnivore" necessary, and let "carnivore" render "four-stomached" impossible; then "lion" may be absolutely denied of "four-stomached."

> All lions are carnivorous;
> No carnivore is four-stomached; therefore
> No four-stomached animal is a lion.

The principle of the other two moods is that "*whatever renders an entity impossible, may be contingently denied of any consequent of that entity.*" Let "negro" render "Hindoo" impossible, and let "Hindoo" be antecedent, either of necessity or of contingency, to "colored"; then "negro" may be contingently denied of "colored."

Both these moods are fitted to produce conclusions of guarded contingency; nor is there any disjunctive mood in the fourth figure whose conclusions are necessarily unguarded. In this respect syllogisms of disjunctive reciprocation resemble those of the separating (or disjunctive) consequents: because, in each case, that same construction of premises which is necessary to produce a negative conclusion, is also fitted to produce a guarded conclusion. In order to either of these results, the premises, after being reduced to the consequent-consequent form by the conversion of one or both premises, must be bound together by a necessitant conception. This may then stand either as predicate of the first (or minor) premise, or as subject of the second (or major) premise; and in every case, while supporting a negative conclusion, it also renders a guarded conclusion possible. Hence there are no moods either of the separating-consequents, or of disjunctive reciprocation, which produce unguarded contingencies only.

13. Nevertheless it is not true that no unguarded contingencies can be inferred by either of these methods. For *if any syllogism in any figure have an unguarded contingent premise, the conclusion must be an unguarded contingency:* because no conclusion can be any better than any premise on which it depends. In all the mood-formulas considered hitherto in this discussion, it has been assumed that no unguarded premise is employed, but that every contingent premise sets forth an absolute possibility either of being or of not being. The inquiry has been, "In what cases do *guarded* premises produce a guarded conclusion, and in what cases do they produce an unguarded conclusion?" The answer to this inquiry is that the conclusion is guarded when there is a connective necessitant; and unguarded when there is not. But, notwithstanding this answer — and whatever be the mood of a syllogism — *the conclusion must be unguarded, if either premise is unguarded.* For the connective necessitant — or "distributed middle term" — adds no new force to the premises, but simply unites them in such a way that there can be no subsidence to the weaker style of sequence.

Syllogisms with an unguarded contingent premise may be neglected, and treated as exceptional, because of their rare occurrence; yet they are possible in any contingent mood, of whatever figure. Let us take the mood *EIO*, in the second figure, and employ for minor premise the unguarded contingency, "an ox may be a carnivore"; inferred because a quadruped may be a carnivore, and an ox is a quadruped. Let us say,

> No four-stomached animal is a carnivore;
> An ox may be a carnivore; therefore
> An ox may not be four-stomached.

Here the conclusion is unguarded, not because of any laxity in the mood, but because of the original deficiency of the minor premise. The connective necessitant "carnivore," found in the major premise, cannot remedy this defect.

A precisely similar result follows in the mood *EIO* of the fourth figure, if we convert the above minor premise, and say,

> No four-stomached animals are carnivores;
> A carnivore may be an ox; therefore
> An ox may not be four-stomached.

14. All the negative moods of the fourth figure employ in the middle places some term and its contradictory; and so immediately appeal to the law of Contradiction. In this particular these moods resemble those of the separating-consequents, and are unlike those of the consequent-consequent: these last do not employ a term and its contradictory, but the very same middle conception twice. Hence we naturally reduce the negative moods of the fourth figure to equivalent moods in the second figure; simply converting the minor premise, we complete the inference by the separating-consequents, without further conversion.

On the other hand, the positive moods of the fourth figure are easily replaced by equivalent moods in the third figure, through a conversion of the major premise; after which the mind can complete its work according to the law of the common-antecedent. But these positive moods of the fourth figure are also easily interpreted by the conversion of both premises, and the use of the consequent-consequent. Nay, syllogisms of the third figure seem, for the most part, to be mentally effected by the consequent-consequent, after conversion of the minor. The law of the common-antecedent does not have so independent and distinctive an operation as that of the separating-consequents. The syllogism of the third figure may be explained as a variation of that of the first, resulting from a comparatively insignificant conversive addition, based on the principle of identity.

15. If these things be so, great prominence should be given to the methods of the consequent-consequent and of the separating-consequents. *The one is the method of conjunctive, the other of disjunctive, syllogizing.* In the one, a given sequence

(major premise) is accompanied by the inferential assertion of its antecedent as the consequent of another sequence (minor premise); and thereupon the consequent of the major premise is asserted: this appeals to that primary use of the law of Reason and Consequent, according to which first the antecedent, and then the consequent, is asserted. In the other, a given sequence (major premise) is accompanied by the inferential denial of its consequent, by reason of this consequent being the contradictory of the consequent of another sequence (minor premise); and thereupon the antecedent of the major premise is denied: this appeals to that secondary use of the law of Antecedent and Consequent, according to which first the consequent, and then the antecedent, is contradicted, or denied. Thus the first and the second figures appeal more directly than the third, and much more directly than the fourth, to the fundamental principle of inference.

CHAPTER XXIV.

THE PURE, OR DOGMATIC, SYLLOGISM.

1. Is the syllogism recognized by modern authorities. Reasons about things as members of logical classes, and on principles relating to such classes. Has the same "figures" with the modal syllogism; but the subject of each proposition always is, and the predicate always may be, a class or part of a class. 2. The first figure arises when the subject of one premise is made the subject, and the predicate of the other premise the predicate, of the conclusion. It is governed by the "Dictum"; and has the moods *Barbara, Celarent, Darii, Ferio*. 3. The second figure arises when the premises have contradictory predicates. Its moods are, *Cesare, Camestres, Festino, Baroko*. 4. The third figure arises when the premises have a common subject. Its moods are, *Darapti, Datisi, Disamis, Felapton, Bokardo, Feriso*. 5. The fourth figure arises when the predicate of one premise is made the subject, and the subject of the other premise, the predicate, of the conclusion. Its moods are, *Bramantip, Dimaris, Camenes, Fesapo, Fresison*. 6. Euler's diagrams. 7. Hamilton's syllogistic notation. 8. Its application to conjectural moods. 9. His multiplication of moods uncalled for. 10. The enthymeme, epicheirema, sorites, and polysyllogism.

1. PURE, or dogmatic, propositions, though in form mere statements of fact, really express laws of necessity and of contingency; and their value arises from their fitness for this use.

That they are not properly factual assertions is evident because the logical classes whose existence they assume are essentially supposed, or hypothetical, entities — creations which the mind makes for the purposes of its thought. When we say, "All men are mortal," or "Some men are wise," the class of which we speak includes all human beings that ever have been or shall be, and, in addition, all that may be supposed or imagined to be. Therefore, as a whole, it is hypothetical in character and use.

Pure, or dogmatic, syllogisms are those composed *exclusively of pure propositions*, and are the only kind recognized by modern authorities. Dr. Thomas Reid, in 1770, in his "Analysis of Aristotle's Organon," passes over the rules for modal syllogisms in silence, saying that in this he follows the example "of all writers in logic for two hundred years back." In 1870, Sir Wm. Hamilton declares that "the modality of propositions and syllogisms ought to be wholly excluded from logic"; and this is the doctrine commonly taught at the present time. The modal syllogism has been "formally expelled from the science," on the ground that it is only a modification of the pure syllogism, and should be interpreted accordingly.

This position is the reverse of truth; the pure syllogism is the secondary mode of thought, and should be interpreted by the modal. At the same time the importance of the pure syllogism cannot be denied. It is the best expression of our ordinary reasonings; and it is the basis of rules which are easily apprehended and applied. Therefore, also, an account of it properly follows discussions in which the laws of the modal syllogism have been explained.

The figures of the dogmatic syllogism are identical with those of the modal, but every proposition used in them asserts something respecting the whole or a part of some logical class considered distributively. Comparing the figures with reference to their immediate operation, they may be named the subordinative, the refutative, the partitive, and the reciprocative. In the first, one truth is inferred as subordinate to another: in the second, a proposition is disproved by an appeal to the law of contradiction: in the third, something is included, positively or negatively, about a part of a class of things; and in the fourth, a mediate predicate (or consequent) becomes conversely — or reciprocally — subject instead of predicate (or antecedent instead of consequent). Thus each figure of syllogizing has a specific operation; though it cannot be said to be confined to this operation as an end.

2. The dogmatic syllogism of the first figure arises when the subject of one premise is made the subject, and the predicate of the other premise, the predicate, of the conclusion. It is governed by the principle known as "Aristotle's Dictum," that whatever is affirmed or denied of a generic *class, distribu-*

tively, may be asserted in the same way of any subordinate classes or individuals. Hence, *as to quality*, (*a*) the major premise may be either affirmative or negative; (*b*) the minor must be affirmative, because it asserts membership in the class; and (*c*) the conclusion must agree in quality with the major. *As to quantity* (*d*) the major must be universal; (*e*) the minor is either universal or particular, according as it asserts that all or a part of a class is contained in the generic class; and (*f*) the conclusion must agree with the minor in quantity.

Combining these rules we find that the major may be A or E; the minor A or I; the conclusion A, E, I, or O; and the following moods are valid, *AAA*, *EAE*, *AII*, *EIO*. These are known by the names Barbara, Celarent, Darii, and Ferio, in each of which the vowels of a mood are presented in their order. To illustrate these moods, we can say,

in Barbara,
All trees are combustible;
All oaks are trees; therefore
All oaks are combustible;—

in Celarent,
No trees are minerals;
All oaks are trees; therefore
No oaks are minerals;—

in Darii,
All oaks are deciduous;
Some trees are oaks; therefore
Some trees are deciduous;—

in Ferio,
No oaks are evergreens;
Some trees are oaks; therefore
Some trees are not evergreens.

Considered mentally, all these moods follow the law of the consequent-consequent. The first two have apodeictic conclusions; the other two, contingent. But these contingent conclusions are guarded against a necessity of the opposite, because the contingency of them is supported by fact. A sequence which takes place occasionally cannot be impossible. Hence these four pure moods agree with those four modal moods, of the first figure, which produce either apodeictic conclusions or guarded contingencies. Or — since every apodeictic statement is guarded against the opposite necessity — we might say that these are essentially the four guarded moods of the consequent-consequent.

We should remark, however, that not every syllogism with a guarded contingent conclusion can be stated in pure propositions. Only inductive, or empirical, contingency, and guarded conclusions from it, can be set forth dogmatically. If the premise be a guarded mathematical, or intuitive, contingency, both premise and conclusion call for modal expression (Chap. XXI.). But, since the contingencies commonly considered are empirical, all ordinary reasoning can be presented in pure syllogisms.

3. Dogmatic syllogisms of the second figure are governed by the axiom that "*if a class have any positive, or any negative, characteristic, universally, then any class or individual which has a contrary characteristic, does not belong to that class* — in other words, does not have the essential nature of that class." According to this (*a*) the major premise may be either affirmative or negative, (*b*) the minor must be opposite in quality to the major, (*c*) the conclusion must be negative. Also, (*d*) the major must be universal, (*e*) the minor may be either universal or particular, and (*f*) the conclusion must agree in quantity with the minor.

Combining these rules, we find that the major may be A or E; if the major is A, the minor must be E or O, but if the major is E, the minor must be A or I; the conclusion must be E or O; and the valid moods are EAE, AEE, EIO, AOO. These are known by the names Cesare, Camestres, Festino and Baroko (or Fakoro). Thus we may say,

in Cesare,
  No sound is visible;
  All color is visible; therefore
  No color is a sound; —

in Camestres,
  All color is visible;
  No sound is visible; therefore
  No sound is a color; —

in Festino,
> No vices are praiseworthy;
> Some habits are praiseworthy; therefore
> Some habits are not vices; —

in Baroko,
> All birds are oviparous;
> Some animals are not oviparous; therefore
> Some animals are not birds.

A slight inspection shows that the above moods really follow the law of the separating-consequents. They are, therefore, of the same nature with those four modal moods in which guarded conclusions are drawn in the second figure.

4. Dogmatic syllogisms in the third figure are best explained by a double law, consisting of two axioms. First, "*if two predicates be affirmed of the same class of things, one of the predications, at least, being universal, then either predicate may be particularly asserted of the other, that is, of the class which the other designates.*" In other words, if one predicate be with all the members of a class, and another either with some or with all, each of these predicates must sometimes be with the other.

This principle calls for two affirmative premises, one of these, at least, being universal; and a particular affirmative conclusion. Therefore the major may be A or I, and the minor A or I, but if either premise be I, the other must be A; the conclusion must be I; and the valid moods are AAI, AII, IAI. These are named Darapti, Datisi, and Disamis; and they are mentally identical with those affirmative moods of the common-antecedent which have guarded conclusions.

The second axiom is that "*if one predicate be denied and another affirmed of the same class of things, one of the predications, at least, being universal, then the predicate denied may be particularly denied of the other, that is, of the class designated by the other.*" For, on the one hand, what is separate from every member of a class, must be separate from what inheres in the members of that class as often as this inherency exists; and, on the other hand, what is separate from a part of a class, must be sometimes separate from that which inheres in every member of that class.

According to this (*a*) the major premise must be negative — it asserts separation from a class of things: (*b*) the minor must be affirmative — it asserts union with that class; and (*c*) the conclusion is negative. Also (*d*) the major must be universal if the minor is particular — otherwise it may be particular; (*e*) the minor must be universal if the major is particular — otherwise it may be particular; and (*f*) the conclusion must be particular. Hence, under the limitations of these rules, the major may be E or O, and the minor A or I; the conclusion must be O; and the valid moods are EAO, OAO, EIO. These are known by the names Felapton, Bokardo (or Dokamo) and Feriso; and they agree with the negative moods of the common antecedent which have guarded conclusions.

Thus dogmatic syllogisms in the third figure have three affirmative moods, and three negative. We can say,

in Darapti,
>All gilding is metallic;
>All gilding shines; therefore
>Some shining things are metallic; —

in Datisi,
>All homicides are cruel;
>Some homicides are lawful; therefore
>Some lawful things are cruel; —

in Disamis,
>Some homicides are lawful;
>All homicides are cruel; therefore
>Some cruel things are lawful; —

in Felapton,
>No moral precept is a material thing;
>All moral precepts are useful; therefore
>Some useful things are not material; —

in Bokardo,
>Some fevers are not infectious;
>All fevers are diseases; therefore
>Some diseases are not infectious; —

in Feriso,
> No punishments are pleasant;
> Some punishments are beneficial; therefore
> Some things beneficial are not pleasant.

5. Dogmatic syllogisms in the fourth figure are most naturally explained by three axiomatic principles.

The first of these pertains to affirmative syllogisms, and is, that "*if a first class be partly, or wholly, included in a second, and the second wholly in a third, then that third is partly comprised in the first.*" If all or some P's be M's, and all M's be S's, then some S's must be P's. According to this the major must be A or I; the minor A; the conclusion I; and we have two valid moods, AAI, IAI. These are called Bramantip and Dimaris; they agree with the guarded moods of affirmative reciprocation (Chap. XXIII.).

We need scarcely say that the P's, the M's, and the S's mentioned above are the classes of things characterized by the major, the middle, and the minor terms respectively; for the major term is the predicate, and the minor the subject, of the conclusion in every syllogism.

The second axiom justifies an universal negative conclusion. It is that "*if a first class (P's) be wholly included in a second (M's), which is wholly excluded from a third (S's), then the third is wholly excluded from the first.*" This principle is the dogmatic expression of the law of absolute disjunctive reciprocation. It supports only one mood, AEE; and this has been named Camenes.

The third axiom justifies particular negative conclusions. "*If a first class (P's) be wholly excluded from a second (M's), which is wholly or partly included in a third (S's), then the third must be partly excluded from the first.*" Evidently this principle supports two moods, EAO, EIO; and these, which are known as Fesapo and Fresison, are substantially those of contingent disjunctive reciprocation.

Thus, in the fourth figure, the dogmatic syllogism has two affirmative and three negative moods. We can say,

in Bramantip,
> All greyhounds are dogs ;
> All dogs are quadrupeds ; therefore
> Some quadrupeds are greyhounds ; —

in Dimaris,
> Some virtuous men are necessitarians ;
> All necessitarians are speculators ; therefore
> Some speculators are virtuous men ; —

in Camenes,
> All ruminants have four stomachs ;
> No four-stomached animal is carnivorous ; therefore
> No carnivore is ruminant ; —

in Fesapo,
> No negro is a Hindoo ;
> All Hindoos are colored ; therefore
> Some colored men are not negroes ; —

in Fresison,
> No moral motivity is an animal impulse ;
> Some animal impulses are principles of action ; therefore
> Some principles of action are not moral motivities.

The fourth figure and its rules employ a mode of thought which is possible in every figure, but which is less called for and less natural in the other figures than in the fourth. Very frequently, in syllogizing, we substantialize and quantify only two of the three terms, and use the third term in an adjective way; that is, we conceive of two classes and of an ascriptive predicate. But, in every pure proposition, the predicate as well as the subject may be quantified and may be taken to represent the whole or part of a class; so that every dogmatic syllogism may be stated as setting forth three logical classes, with certain relations between them of inclusion and of exclusion.

6. In connection with this mode of conception, diagrams have been employed to illustrate the moods of pure syllogisms by means of plane figures. In Euler's "Letters to a German Princess," a circle is used to symbolize the class of things designated by a term. The universal affirmative proposition is indicated by completely enclosing a subject-circle within a predicate-circle; the particular affirmative by a subject-circle

which is partly included in the predicate-circle; the universal negative by a subject-circle which is completely excluded from the predicate-circle; and the particular negative by the same diagram as the particular affirmative, but with the understanding that the subject-circle is partly excluded from the predicate-circle.

These symbols do not serve so well for particular as for universal assertions; because they show to the eye only that "some" are, or are not, without adding that "perhaps all" are, or are not. This, however, being understood, every mood in every figure may be represented geometrically. For example, in the fourth figure, the mood Bramantip, which asserts that

All P is M;
All M is S; therefore
Some S is P—

assumes visible shape when we indicate the major premise by enclosing a first circle in a second, and the minor by enclosing that second in a third. For then these circles show plainly that some of S must be P. This same diagram may set forth Barbara of the first figure. Only, for this end, the intermediate-circle should be drawn first, the outer one second, the innermost last; and the letters P and S should exchange places.

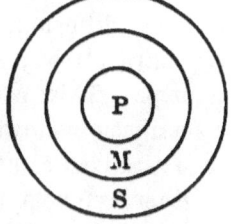

Dimaris, in the fourth figure, is diagrammed by first drawing the circle P so as to intersect the circle M; this expresses the major premise; then, by circumscribing M with another circle S, we express the minor premise. Thus it is made to appear that some S's must be P's.

This same diagram, without any change, illustrates Disamis, of the third figure, if we first draw M, then P, and then S. For then we can say,

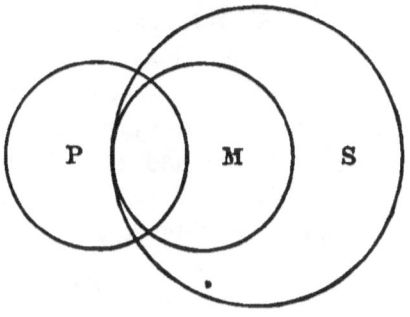

Some *M* is *P*;
All *M* is *S*; therefore
Some *S* is *P*.

Again, Camenes, of the fourth figure, is explained, if we draw one circle within a second, to represent the inclusion asserted in the major premise; and then a third circle outside of the second, to represent the exclusion asserted in the minor premise. Plainly no *S*'s are *P*'s.

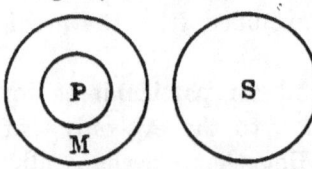

This diagram, without any change, serves for Camestres of the second figure; which indicates that Camestres and Camenes differ very little.

Should we neglect the lettering, one diagram will represent Celarent, Cesare, Camestres, and Camenes; one, also, will suffice for Darii, Disamis, Datisi, and Dimaris; and another, varying but slightly from this, will express Darapti. One will be sufficient for Ferio, Festino, Feriso, and Fresison; and another, differing little from this, for Felapton and Fesapo. We have already seen that one serves for Barbara and Bramantip. This community in symbolization suggests, what investigation shows to be the fact, that different sets of moods have a radical sameness of nature and operation.

Other symbols than plane figures have been used for the ocular illustration of propositions and syllogisms. The German logician, Lambert, and after him Sir William Hamilton, employed parallel lines; but this plan proved difficult of application in the subordinate figures.

7. Then Sir William Hamilton devised an excellent notation, in which propositions (not terms) are represented by horizontal lines thickened at the subject end, and sharpened at the predicate end; terms are indicated by letters; the distribution of a term by the addition of a colon; and its non-distribution by a comma. Affirmation is expressed simply by the line; negation by the line with a dash across its centre.

P: ▬▬▬▬▬ ,S

means "All *P*'s are (some) *S*'s," and is equivalent to *A*.

P: ━━━━━━+━━━━━━ :S

means "No P is (any) S," and is equivalent to E. In stating syllogisms, we first place the middle term (that is, a letter indicating it) in the middle, the minor on the right, and the major on the left; then insert the proper marks between each extreme and the middle; and, finally, draw a long line, either above or below, to indicate the conclusion.

Barbara is expressed thus:

P, ━━━━━━━━ :M, ━━━━━━━━━━ :S
P. ━━━━━━━━━━━━━━━━━━━━━━━ :S

For this, in accordance with the explanations given above, reads,

All M are P;
All S are M; therefore
All S are P.

As the letters at the extremities of the conclusion-line are always repeated from the premises and can be easily understood, they may as well be omitted. So also may the marks of quantity except when a term loses in the conclusion the distribution which it had in the premise. In the foregoing symbolization, the conclusion-line would be sufficient alone, but in that for Bramantip a new sign must be used at the predicate end of the line; thus,

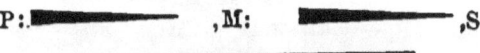

For this now reads, according to the order of the fourth figure,

All P are M;
All M are S; therefore
Some S are (some) P.

Festino, of the second figure, is written thus:

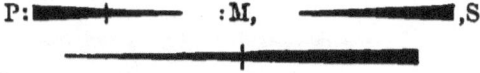

for this reads,

No P are (any) M;
Some S are (some) M; therefore
Some S are not (any) P.

According to this admirable notation, the nineteen dogmatic moods are represented as follows:

272 THE MODALIST. [Chap. XXIV.

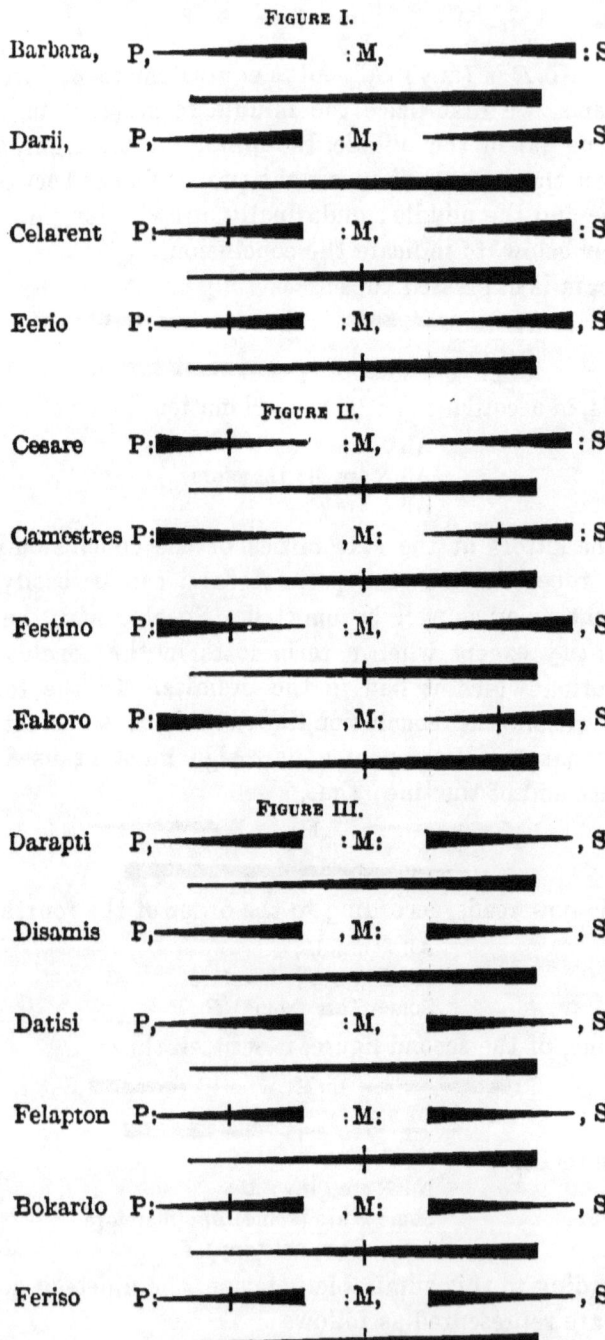

CHAP. XXIV.] *THE PURE, OR DOGMATIC, SYLLOGISM.* 273

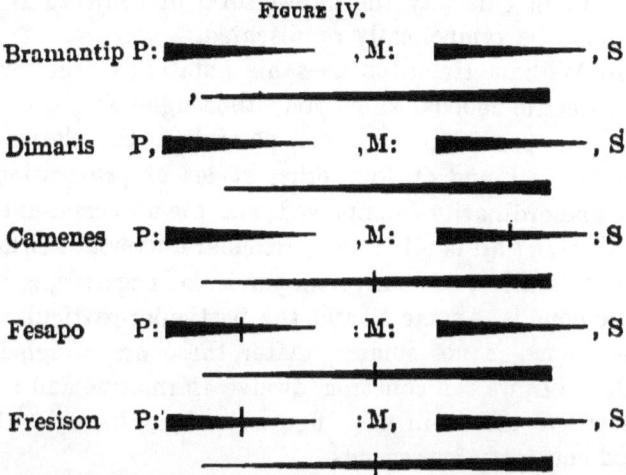

FIGURE IV.

8. The twelve contingent moods with unguarded conclusions, which the doctrine of modality adds to the nineteen moods which are capable of dogmatic expression, may be set forth by the above symbolization provided a very elongated triangle take the place of the thick tapering line in representing the conclusion. For example, in the first figure we would have,

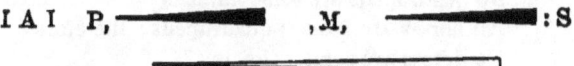

I A I P,   ,M,   :S

For we can say,

> Some uncivilized people are treacherous;
> The Hottentots are uncivilized; therefore
> A Hottentot may be treacherous.

In the second figure we might have

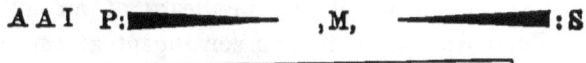

A A I P: ,M,   :S

and so on with the rest of the twelve moods. The commas in the conclusions of these moods indicate contingency, not particular quantity; and they might be omitted; for every conclusion is contingent.

Moreover, the unguarded contingent premises, which we sometimes use, might also be indicated by the elongated triangle; and in this way the purest form of conjectural syllogizing could be symbolically represented.

9. Sir William Hamilton uses his notation to set forth all those dogmatic moods which his "thorough-going quantification of the predicate" renders possible. He claims that, besides A, E, I, and O, four other styles of propositions are often — and ordinarily — employed; viz., the universal-universal affirmative, U, "all is all"; the particular-universal affirmative, Y, "some is all"; the universal-particular negative, η, "all is not — or none is — some"; and the particular-particular negative, ω, "some is not some." After these are mingled with A, E, I, and O, we can construct twelve affirmative and twenty-four negative moods in each figure; and so have in all one hundred and forty-four moods.

But none of the moods added by Hamilton's scheme to those previously recognized, should be numbered among regular logical forms. If either η or ω be used as a premise, the syllogism is abortive because of a negative conclusion with undistributed predicate. For example, combining η and A in the first figure we have the mood $\eta A\eta$; thus,

> No quadrupeds are some animals;
> All horses are (some) quadrupeds; therefore
> No horses are some animals.

This conclusion fails to characterize either positively or negatively; and so does every conclusion dependent on η or ω.

Then the use of U or Y as a premise is an extraordinary occurrence, and should be treated as exceptional. Ordinarily — always, except after certain necessary mental conversions — we neglect the quantity of the predicate of an affirmative premise. For both apodeictic and contingent affirmations are fully expressed without such quantification.

Hamilton's multiplication of moods sprang from the theory that the essential aim of syllogizing is to show how far one logical class includes, or excludes, another; whereas it is to show whether, and in what way, the subject of the conclusion, as antecedent, may be related to the predicate of the conclusion, as consequent.

10. In connection with the pure syllogism certain terms may be defined, which are used chiefly to describe contracted modes of argument.

When only one premise of a syllogism is expressed, the other being understood, the argument is called an enthymeme. This is of the *first order*, if the major premise be omitted; as when we say, " Comets are subject to gravitation ; because they move in elliptic orbits " : and it is of the *second order*, when the minor premise is suppressed ; as when we say, "Comets obey the law of gravitation ; because all bodies which move in elliptic orbits, are subject to that law."

When an enthymeme is used as the premise of a syllogism, the argument is styled an epicheirema ; and an epicheirema is *double* when both of its premises are enthymematic ; and *single*, if only one of them have that character. The following is a single epicheirema ; in which the minor premise is an enthymeme of the first order :

>All vice is odious ;
>But avarice is a vice ; for it depraves men ; therefore
>Avarice is odious.

The sorites, or chain-argument (*die schluss-kette*), is the abbreviated statement of a series of syllogisms formed immediately according to the consequent-consequent, and belonging, therefore, to the first figure. It consists of a succession of catenated sequences ; and terminates with the conclusion that the consequent (or predicate) of the last proposition, follows the antecedent (or subject) of the first. In addition to the first premise and the conclusion, the sorites has as many propositions as it contains middle terms.

The following is a chain of reasoning, quoted by Sir William Hamilton from Seneca :

>He who is prudent is temperate ;
>He who is temperate is constant ;
>He who is constant is unperturbed ;
>He who is unperturbed is without sorrow ;
>He who is without sorrow is happy ; therefore
>The prudent man is happy.

This argumentation, following the natural, not the dialectic, order of reasoning, first makes what is to be the subject of the conclusion antecedent to a middle term; then it makes that middle term antecedent to a second; and that second middle term antecedent to a third; and so on, till the last premise makes the last middle term antecedent to that which is to be the predicate of the conclusion: whereupon the conclusion is asserted. This reasoning is easily resolved into the following syllogisms of the first figure; in each of which the minor premise is written first:

(*a*) He who is prudent is temperate;
He who is temperate is constant; therefore
He who is prudent is constant.

(*b*) He who is prudent is constant;
He who is constant is unperturbed; therefore
He who is prudent is unperturbed.

(*c*) He who is prudent is unperturbed;
He who is unperturbed is without sorrow; therefore
He who is prudent is without sorrow.

(*d*) He who is prudent is without sorrow;
He who is without sorrow is happy; therefore
The prudent man is happy.

The above order of syllogisms — as well as that of the premises in each syllogism — is called the *progressive*; because, beginning with what is to be the antecedent of the conclusion, it follows the links of the chain (that is, the middle terms) from the antecedent-end to the consequent-end. It is distinguished from the *regressive* order; according to which single syllogisms of the first figure are commonly stated; and which begins with the consequent-end, and therefore with the major premise.

The first of the five premises in the foregoing sorites may be termed the minor; because it contains the minor term, the subject of the conclusion: the last of the five, the major; because it contains the major term: the rest of the five are neither minor nor major, but simply middle premises; which, however, act as major premises when the series of syllogisms is developed.

The rules governing the sorites are entirely analogous to those of the first figure; and are as follows:

(*a*) Only the minor premise may be particular (or contingent); the rest must be apodeictic.

(*b*) Only the major premise may be negative; the rest must be affirmative.

(*c*) The conclusion agrees in modality with the minor premise; and in quality with the major.

But the first of these rules is necessary only when the conclusion is to be guarded against a necessity of the opposite; a series of compounded probabilities may be set forth in a chain-argument, and may have an unguarded conclusion.

While the progressive is the natural order of thought for a sorites, the regressive order, also, may be used. We can say:

>He who is without sorrow is happy;
>He who is unperturbed is without sorrow;
>He who is constant is unperturbed;
>He who is temperate is constant;
>He who is prudent is temperate; therefore
>The prudent man is happy.

The argumentation, as thus stated, may be developed into the following syllogisms; in each of which the major premise is put before the minor:

>(*a*) He who is without sorrow is happy;
>He who is unperturbed is without sorrow; therefore
>He who is unperturbed is happy.
>
>(*b*) He who is unperturbed is happy;
>He who is constant is unperturbed; therefore
>He who is constant is happy.
>
>(*c*) He who is constant is happy;
>He who is temperate is constant; therefore
>He who is temperate is happy.
>
>(*d*) He who is temperate is happy;
>He who is prudent is temperate; therefore
>The prudent man is happy.

In the above syllogizing we begin with the major term, and carry it backwards through the series, till it is connected with the minor; whereas the other mode of syllogizing began with the minor term, and carried it forwards through the series, till it was connected with the major.

Rudolphus Goclenius, of Marburg, a distinguished professor of logic in the seventeenth century, called attention to the regressive sorites, and to its conformity with ordinary syllogistic formulas. Hence this form of statement has been named the Goclenian sorites.

When the conclusion of a first syllogism becomes the premise of a second syllogism, the relation between the arguments is indicated by calling the first a *prosyllogism*, and the second an *episyllogism*. This is especially the case when the conclusion of the first syllogism is the only premise required for the second syllogism; the other having been already provided. When a sorites is developed into syllogisms, every one of these, except the last, is a prosyllogism, with reference to the succeeding syllogism; and every one except the first, is an episyllogism, with reference to the one preceding it. The whole series of syllogisms has also been called, sometimes, a *polysyllogism*.

CHAPTER XXV.

REDUCTION OF SYLLOGISMS.

1. The four figures as related to each other, and to the law of the consequent-consequent. 2. All logicians have followed the methods of Aristotle in his syllogistic "reductions." A new, and natural, method proposed. 3. The ordinary reductions are often artificial and indirect. The mnemonic mood-names. The significance of their vowels. 4. The significance of their initial consonants, and of the letters *s*, *p*, *m*, and *k*. The expressions "per accidens" and "per se." 5. Reduction "per impossibile," or "per contradictionem." 6. Exercises in syllogistic construction and reduction. 7. Also in forming and reducing unguarded syllogisms.

1.[1] The law of the first figure is that "*if a first thing* (minor term) *be antecedent to a second* (middle term), *and this second to a third* (major term), *then also the first is antecedent to the third.*" We call this "the consequent-consequent"; because the third thing, as being consequent to the second, is consequent to the first, and so becomes a consequent-consequent. It might also be named the consequent-consequence; because the consequence — or sequence — in which the third thing follows the first, is consequent upon the combination of the two premises, or prior sequences.

The law of the second figure is that *if a first thing* (minor term) *have a second thing as either a positive or a negative consequent* (middle term), *and a third thing* (major term) *be necessarily followed by the contradictory of that consequent* (middle term), *then the first thing is antecedent of denial to the third.* If

---

[1] In these discussions it must be remembered that the major term is always the predicate of the conclusion, and the minor term the subject of the conclusion; and then that *the major premise is always that which contains the major term, and the minor premise is that which contains the minor term.* The order in which premises are stated determines nothing as to either figure or mood.

A be followed by B, and C necessarily by not-B; or if A be followed by not-B, and C necessarily by B; in either case, A is followed by not-C. This is the principle of "the separating-consequents"; it might also be named the law of consequent contradiction, or denial. It is constituted by uniting to the syllogistic law of the consequent-consequent the conversional law of the denied-consequent. For when the sequence which must have the necessary consequent, that is, the major premise, is converted by the denied-consequent, the argument immediately falls into the first figure. Thus the syllogism,

> No Germans are black (minor premise);
> All negroes are black (major premise); therefore
> No Germans are negroes,

by the conversion of the major, becomes,

> A German is not black;
> He who is not black is not a negro; therefore
> No German is a negro.

But this conversion, which is required by the separating-consequents, is in most cases potential — not actual; for, by an acquired habit, we can immediately deny the one antecedent of the other. Nor is the cogency of the argument increased by the change from the second figure to the first.

The law of the third figure is that "*if two consequents have a common antecedent* (middle term), *either may be contingently asserted of the other.*" This law is accounted for by combining with the law of the consequent-consequent the conversional law either of the asserted-consequent or of the retained-necessitant; the latter of these being a specific mode of the former. But while the second figure implicitly converts the major premise, the third figure calls for a converted minor. Thus the syllogism (of the mood Datisi),

> All homicides are cruel;
> Some homicides are lawful; therefore
> Some lawful things are cruel,

assumes the first figure on the conversion of the minor premise by the asserted-consequent.

But, in the syllogism (in Disamis),

> Some homicides are lawful;
> All homicides are cruel; therefore
> Some cruel things are lawful,

we must convert by the retained necessitant; thus,

> Some homicides are lawful;
> Some cruel things are all the homicides; therefore
> Some cruel things are lawful.

So, also, the following (in the mood Bokardo)

Some writers on logic are not clear thinkers;
But all writers on logic profess to teach the laws of thought; therefore
Some who profess to teach the laws of thought, are not clear thinkers,

on converting the minor, becomes

Some writers on logic are not clear thinkers;
But some who profess to teach the laws of thought, are all the writers on logic; therefore
Some who profess to teach the laws of thought, are not clear thinkers.

In the above reductions — of Disamis and Bokardo — the retained necessitant is necessary, in order that the conclusion may not become unguarded. In Darapti and Felapton, also, the minor premise may be converted by the retained necessitant; but this is not needed for a guarded conclusion; the asserted consequent is sufficient; though the conclusion thus obtained in these moods is not so strong as that which would follow the retained necessitant.

The operation of the third figure is closely allied to that of the first. Its conversions suggest themselves more easily than those of the second figure; these latter are driven back from prominence by the action of the principle of contradiction. In the course of discussion an argument given in the third figure is often spontaneously restated in the first; this seldom happens with an argument in the second figure. Yet the third figure, as well as the second, has an independent operation.

The fourth figure has less independence than either the second or the third; and also less unity of principle. It has three laws, one for conjunctive, and two for disjunctive, reciprocation (Chap. XXIV.).

Though this figure may at times operate independently, probably most arguments in it are effected by a mental conversion which reduces the syllogism to one of the other figures. Two conversions reduce any mood to the first figure; but the negative moods naturally fall into the second figure by converting the minor premise, and the positive moods into the third by converting the major (Chap. XXIII.).

2. Aristotle discusses only three figures, probably discarding the fourth as merely an irregular derivative from the others. For, if we take either Bramantip, Dimaris, or Camenes; and — what is merely a matter of order — transpose the premises of the mood; we obtain a conclusion in the *first* figure, of which the conclusion in the fourth figure is the converse: if we take either Camenes, Fesapo, or Fresison (the negative moods); and convert only the minor premise; we obtain the very same conclusion in the *second* figure, as in the fourth: and, if we take either Bramantip or Dimaris (the affirmative moods); and convert the major premise; we obtain the very same conclusion in the *third* figure, as in the fourth. Moreover, every one of these reductions improves the statement of the argument. This fact shows the inferiority of the fourth figure; and explains, though it does not wholly justify, the neglect of that style of syllogizing, by Aristotle and others.

The fourth figure has been ascribed to Galen, who lived in the second century of the Christian Era, but, according to Sir William Hamilton, it is first mentioned by Averroes, who lived in the twelfth century.

Aristotle proved the validity of reasonings in the second and third figures by showing that they are equivalent to reasonings in the first; and all subsequent logicians have felt bound to reduce the moods of each subordinate figure to equivalent moods in the first figure.

This reduction would present no difficulty, if it were only borne in mind that the universal, or apodeictic, affirmative may be converted in any one of three ways; either by *the asserted-consequent* (per accidens), or by *the retained-necessitant* (per differentiam), or by *the denied-consequent* (per contradictionem). We must also remember that the apodeictic negative can be converted by the *denied-consequent;* the contingent affirmative, by the *asserted-consequent;* and that the particular negative is *not to be converted* at all. With these rules any mood in a subordinate figure can be immediately reduced to the first figure, without any change in the order of its premises, and without any alteration whatever in the conclusion.

In every mood of the second figure we simply *convert the major by the denied-consequent.* In this way, for example, Baroko,

> All birds are oviparous;
> Some animals are not oviparous; therefore
> Some animals are not birds,

becomes Ferio,

> No non-oviparous are birds;
> Some animals are non-oviparous; therefore
> Some animals are not birds.

In the third figure we *convert the minor by the asserted-consequent; or by the retained-necessitant, if the case so require.* For when only the minor premise is apodeictic, the full distributive force of its antecedent must be retained: otherwise the conclusion would not be a guarded contingency. This has been exemplified above (Chap. XXIV.).

The fourth figure is immediately reduced by *converting both premises.* For example, the mood Camenes,

> All ruminants are four-stomached;
> No four-stomached animal is a carnivore; therefore
> No carnivore is a ruminant,

becomes, by the retained-necessitant and by the denied-consequent, the mood *YEE*, in the first figure.

> Some four-stomached are all the ruminant;
> No carnivore is four-stomached; therefore
> No carnivore is a ruminant.

In like manner, every mood in this figure may be reduced by two conversions.

Reductions, after the manner now proposed and explained, accord with the philosophy of the syllogism and its figures; and should satisfy those who recognize that the universal affirmative can be converted in three ways.

3. But, till recent times, logicians have subjected this style of proposition only to one method of conversion; that is, to conversion *"per accidens,"* or "by limitation," or, more absolutely speaking, by the asserted-consequent. Under this restriction the reduction of syllogisms was attended with difficulties. Indeed, seven, out of the fifteen moods of the subordinate figures, proved irreducible. It was found possible, however, to reach the conclusions of these moods through the first figure by more or less "indirect" reductions — in other words, by processes which involve the aid of supplementary devices.

For the correct performance of the reductions thus brought into use certain rules were found necessary; and these have been compactly preserved for us in the famous lines of Petrus Hispanus, who lived in the fourteenth century, and who is known also as Pope John XXII.:

> Barbara Celarent Darii Ferioque prioris;
> Cesare Camestres Festino Baroko secundæ;
> Darapti Disamis Datisi Felapton Bokardo
> Feriso sunt sex modi Tertiæ. Quarta insuper addit
> Bramantip Dimaris Camenes Fesapo Fresison.

This list includes all dogmatic moods which can be constructed with *A, E, I,* and *O*; and therefore, modally speaking, all ordinary moods which are either apodeictic or of a guarded contingency. Moods with an unguarded conclusion were rejected because they cannot be expressed dogmatically; and are, also, in a certain sense, inconclusive.

The ingenuity of the mnemonic lines lies in the names given to the moods.

The character of each mood as to the quality and quantity, or modality, of its propositions, is indicated by the three vowels which its name contains. In Darii, for example, of the first figure, or in Datisi, of the third, the major must be an universal affirmative; the minor, a particular affirmative; and the conclusion, a particular affirmative.

4. Again, the names of the four moods of the first figure begin with the first four consonants of the alphabet; and the name of every mood in each of the other figures also, begins with one or other of these letters. This informs us to what mood in the first figure any mood in a subordinate figure is to be reduced. Camestres is to be reduced to Celarent; Felapton to Ferio.

Further, while only the initial consonants are significant in the first figure, in the other figures practical directions are given by means of the letters s, p, m, and k. S, which is inserted only after E or I, signifies that the proposition which it follows is to be converted "simply," or without change in quality or quantity: in Fresison both major and minor are to be converted simply.

Only E and I can be converted in this way. This does not mean, however, that both these conversions take place on the same principle; for the conversion of E follows the denied-consequent, while that of I follows the asserted-consequent. Moreover, the operation of these laws is "simple" only in the sense of being intelligible. In the one case the absolute denial of an antecedent follows the contradiction of its necessary consequent; in the other an antecedent of contingency is contingently asserted, on its consequent being assumed to exist.

The explanation of the conversion of E by the mutual exclusion of two classes, and of I by the reciprocity of partial inclusion between two classes, is also simple; but only superficially. Nothing is truly and philosophically simple which cannot be understood without further explanation. The mutual exclusions and inclusions of logical classes are only a mental device for vividly expressing reciprocations of impossible and of contingent sequence.

The letter p, when it follows the vowel A, signifies that the apodeictic affirmative is to be converted "*per accidens*"; or, as we would prefer to say, by the asserted-consequent, and on the principle of the contained-conditions (Chap. XXI.). This conversion is of the same nature with that of the particular, or contingent, affirmative. But it differs from the latter in that the antecedent of the convertend, which contains a necessitating condition of the consequent, after becoming predicate of the converse, drops its necessitant and distributant force; although this might have been retained. We convert "man must die" into "what dies may be a man," when, were the full force of the convertend retained, we would say, "Only what dies — or is mortal — can be human."

Conversion *per accidens* was formerly confined to the apodeictic affirmative; but it is applicable to the universal negative as well. "No fishes are viviparous," yields "some animals not viviparous are fishes"; and, for the same reason — that is, because of the law of contained-conditions in its negative operation (Chap. XXI.), the particular negative may be converted in the same way. "A fish may not be a marine animal," yields "what is not a marine animal may be a fish."

In saying that *A* is converted *"per accidens,"* the Latin logicians contrasted the modality of the converse with that of the convertend. In the latter, consequent follows antecedent *universally* (καθόλου), *necessarily* (ἐξ ἀνάγκης), or *per se* (καθ' αὑτό); in the former, *particularly* (ἐν μέρει), *contingently* (ἐνδεχομένως), or *per accidens* (κατὰ συμβεβηκός) — in each case the same mode of sequence being viewable in either of three ways. "Universally" and *"per se"* are secondary modes of stating necessity; "particularly" and *"per accidens"* are secondary expressions of contingency. Man *"per se,"* or necessarily, is a mortal and a terrestrial being; but *"per accidens,"* or contingently, he is a wise being, or an Asiatic.

In this connection, *"per se"* does not mean "by a nature, or essence, alone," but *"by a nature under any circumstances whatever"*; and *"per accidens"* does not mean "without the nature," but rather *"by means of the nature under these circumstances; or under those."* (*Vide* Arist., "Analyt. Post.," I. 4.)

The letter "*m*," in the mnemonic names, calls for a mutation, or transposition, of the premises; so that the major becomes minor, and the minor, major. This was found unavoidable in five moods. But the major premise furnishes the predicate, and the minor premise the subject, of the conclusion; therefore the transposed premises, though they bring the argument into the first figure, do not produce the original conclusion, but only a conclusion from which the original can be obtained by conversion. For example, the following, in Camestres,

> All color is visible;
> No sound is visible; therefore
> No sound is a color,

becomes, in Celarent,

> Nothing visible is a sound;
> All color is visible; therefore
> No color is a sound;

in which the conclusion is convertible into the original conclusion. Hence the final s, in the mood-name.

Aristotle himself reduces Camestres in this way; and indeed all the rules of Hispanus simply embody the methods of Aristotle. ("Analyt. Prior.," I. 5.)

Here it should be noticed that p in Bramantip does not indicate the conversion of I, but of A so as to produce I. For the transposed premises produce a syllogism in Barbara.

5. Finally, "k" signifies that the mood is to be reduced "*per impossibile*," that is, by an appeal to the principle of contradiction. Two moods, Baroko and Bokardo, defeated all attempts to reduce them, either directly or with mutation of premises. If, however, we substitute, for the premise which "k" follows, the contradictory of the conclusion, and retain the other premise, we can obtain a syllogism of the first figure. The conclusion of this syllogism will be the contradictory of the suppressed premise; and therefore, as contradicting what was laid down at the beginning, cannot be true. But, as this false conclusion results simply from using the contradictory of the conclusion as a premise, *that contradictory must be false; and the original conclusion, which it contradicts, must be true.*

In Baroko we say,

> All birds are oviparous;
> Some animals are not oviparous; therefore
> Some animals are not birds.

Substituting for the minor the contradictory of the conclusion, we have

> All birds are oviparous;
> All animals are birds; therefore
> All animals are oviparous.

But this conclusion is the contradictory of the original minor, and must be false; therefore the substituted premise must be false, and its contradictory, the original conclusion, must be true.

A precisely similar proof of Bokardo is obtained by substituting the contradictory of the conclusion for the major premise.

Reductio per impossibile may be effected in any mood of the second figure in the same way as in Baroko; and in any mood of the third figure in the same way as in Bokardo. It was specially assigned to these moods in the belief that they could not be reduced in any other way. But Baroko may be reduced directly (according to the new method of reduction, already explained), if we convert the major by the denied consequent; and Bokardo, if we convert the minor by the retained necessitant.

6. Such are the rules of reduction. To promote familiarity with them, and with the laws of syllogizing generally, a few exercises in connection with some such table of terms as the following, will be found helpful.

Figure I.

Moods.	Major.	Middle.	Minor.
Barbara	Elastic.	Gas.	Oxygen.
Celarent	Faultless.	Finite.	Angel.
Darii	Laudable.	Virtues.	Habits.
Ferio	Reprehensible.	Virtues.	Habits.

Figure II.

Cesare	Material.	Free-will.	Spirit.
Camestres	Color.	Visible.	Sound.
Festino	Vices.	Praiseworthy.	Actions.
Baroko	Birds.	Oviparous.	Bipeds.

Figure III.

Darapti	Metallic.	Gilding.	Shines.
Disamis	Laudable.	Homicides.	Cruel.
Datisi	Cruel.	Homicides.	Lawful.
Felapton	Moral.	Material.	Extended.
Bokardo	Wise.	Men.	Rational.
Feriso	Advantageous.	Dishonesty.	Tempting.

Figure IV.

Moods.	Major.	Middle.	Minor.
Bramantip	Greyhounds.	Dogs.	Quadrupeds.
Dimaris	Virtuous.	Necessarians.	Speculators.
Camenes	Ruminant.	Four-stomached.	Carnivore.
Fesapo	Negroes.	Hindoos.	Colored.
Fresison	Moral Principle.	Animal Impulse.	Principle of Action.

Let a syllogism be constructed in every mood of each figure with the terms given above; and then let each argument in a subordinate mood be reduced to the first figure. In this latter work the rules of Hispanus may be employed first; and then that simpler method which has been proposed, and which merely converts the major premise in the second figure; the minor, in the third; and both premises, in the fourth.

7. *After all this, the construction of the unguarded, or conjectural, moods, and their reduction (according to the new method) will present no difficulty; and may be exemplified in connection with the following table of terms:

Figure I.

Moods.	Major.	Middle.	Minor.
IAI	Fatal.	Accidents.	Railroad Collision.
III	Over-indulged.	Pet.	Dog.
OAO	Fatal.	Diseases.	Fever.
OIO	Profitable.	Speculation.	Investment.

Figure II.

AAI	Horse.	Animal.	Quadruped.
AII	Serpent.	Reptile.	Venomous.
IAI	Venomous.	Reptile.	Serpent.
III	Metal.	Hard.	Mineral.

Figure III.

III	Metal.	Hard.	Mineral.
OIO	Metal.	Hard.	Mineral.

Figure IV.

AII	Serpents.	Reptiles.	Venomous.
III	Intelligent.	Man.	Unprincipled.

These unguarded modal syllogisms have hitherto been overlooked. But they express a conjectural kind of reasoning which is not uncommon.

CHAPTER XXVI.

FALLACIES.

Paralogisms in Separate Inference.

1. Our simple and immediate perceptions are reliable. Error arises only in connection with the complex, and the inferential. Fallacy, paralogism, sophism, defined. Truth and falsity are either logical or real. 2. In correct argument a conclusion really false, shows falsity in the premises; but a conclusion really true may follow from false premises correctly. In fallacious reasoning premises and conclusion do not in any way involve each other. 3. Eight forms of inference, each of which has its own paralogisms. 4. And each of which may be expressed syllogistically. 5. A comprehensive enumeration of fallacies. 6. The hypothetical syllogism (and its fallacies) discussed. 7. Disjunctive syllogisms. 8. Relational syllogisms. 9. Problematic syllogisms. 10. The paradigmatic syllogism. 11. Principiative, and inductive, syllogisms. 12. The applicative syllogism. 13. The Aristotelian, or catenate, syllogism.

1. Perceptions, absolutely simple and immediate, are exempt from error. But, in a complex apprehension, or in a process of inference, the mind can suppose some element to be present when it is absent, or absent when it is present; and can wrongly connect a consequent with an antecedent. In these ways error may arise, in any finite intellect. The liability to error, however, is greater in some than in others; and it is so far unnatural to a sound intellect that it can be guarded against, and often entirely obviated, by care and circumspection.

An act or process of argument may have the appearance of being conclusive, without being really so. In that case, in token of its fitness to mislead, it is called a fallacy; simply as a deviation from the laws of right reasoning, it is named a paralogism; and, as employed with the intention to deceive, it is termed a sophism.

The conclusion asserted in a fallacy is often said to be false, no matter whether it state truth or not. This signifies merely that it is falsely asserted to follow from the premises. In the syllogism,

> Good men are sincere;
> The apostles were sincere; therefore
> They were good men,

all the propositions are true; but the conclusion is logically false.

In like manner, that is often called a true conclusion which necessarily follows from the premises; whether it be true in itself or not. In this sense the syllogism

> All Orientals speak Arabic;
> All the Chinese are Orientals; therefore
> All the Chinese speak Arabic,

has a true conclusion; though the proposition in itself is false.

2. In every correct argument, if nothing be falsely assumed, the conclusion must be true in itself, as well as logically. Therefore, if a conclusion be logically true, but really false, one or other, or both, of the premises must be false. What is false can be correctly inferred only from what is false.

But we cannot say that, if a conclusion has been correctly drawn, the premises must be true. A consequent may be inferred, not from one antecedent only, but from many; and among these may be those purely imaginary, and unreal. We might say,

> All stones are moral beings;
> All men are stones; therefore
> All men are moral beings;

and this syllogizing would be correct.

In right reasonings, therefore, if the premises be true, the conclusion is true, and if the conclusion be false, the premises are false, or at least one of them; but we cannot say that if the premises be false, the conclusion is false; nor that if the conclusion be true, the premises are true. On the other hand, in fallacious reasoning, there being no true sequence, the conclusion asserted may be either false or true, no matter what the character of the premises may be; and the premises, likewise, either false or true, no matter of what character the conclusion may be.

3. Fallacies may occur *in connection with every form of inference; and should be discussed in connection with each separately.* Especially we must avoid that confusion which attends the theory, that there are only two or three species of inference; for no confusion is more perplexing than that of a false simplification. Logical sequence has a variety of modes, all subject to the generic law of reason and consequent, but each of which has its own rules, and is affected by fallacy in its own way.

In order to judge wisely respecting sequence, and error in sequence, we must distinguish the translative, the disjunctive, the relational, the problematic, the paradigmatic, the principiative, the applicative, and the catenate, inferences.

4. In each of these modes of sequence, with one or two partial exceptions, the antecedent is complex, and may be divided into two parts; and so the process may be given a syllogistic shape — in other words, may be expressed by two premises and a conclusion. For this is what we mean at present by a syllogism. All the above-mentioned modes of inference have been syllogized, either anciently or in modern times, except, perhaps, the problematic, that is, the immediate inference of some consequent as probable, or as contingent, or as possible. But probable sequence may be analyzed into

(1) a fact, or antecedent, originating certain chances, — as that a ball is to be drawn from a bag in which there are 30 white and 70 black balls;

(2) a determination of the ratio of the chances as being 3 to 7, or 3 out of 10, or $\frac{3}{10}$, in favor of a white ball; and

(3) a conclusion, with the probability of $\frac{3}{10}$, that a white ball will be drawn. Thus we obtain two premises and a conclusion.

So, in every contingent judgment, there is

(1) an antecedent fact originating chances;

(2) the perception that an appreciable proportion of these favor a specific consequent; and

(3) the corresponding assertion of that consequent.

Syllogizing after this fashion may not be ordinarily needful for the critical understanding of problematic sequence; but it will be found helpful in most cases of the "calculation of probability," and in the analysis of certain fallacies.

The inference of possibility, which is the simplest mode of problematic sequence, presents one of the partial exceptions mentioned above. It arises when A either is, or contains, a condition of B. In the former case we say,

> The being is rational; therefore
> He may — or may not — be wise.

Here the antecedent is properly expressed by one premise, or one term. But when the antecedent *contains* the condition, we might say,

> (1) Man involves rationality;
> (2) Rationality conditions wisdom; therefore
> (3) Man may — or may not — be wise.

This is a syllogism; indeed, as generalized, it is a conjectural catenate syllogism.

Relational sequences intuitively and orthologically assert the existence of some specific relation or *relatum* as necessarily connected with some antecedent; they are the immediate perceptions of the metaphysical and mathematical connections of things, and of things as in these connections. We call them relational only *par eminence;* and because they are founded on relations which are specific, and not on those which are common, or universal. When they have very simple antecedents they present a second partial exception to the rule that every mode of inference may be given appropriate syllogistic expression; for example,

> There is a body; therefore
> There is occupied space; —
>
> There is a power; therefore
> There is a substance in which it resides; —
>
> There is an action; therefore
> There is an agent.

In such cases the antecedents are too simple for analysis. But if these simple antecedents be embodied in others more complex, we can use analysis and say,

(1) The stone is a body ;
(2) As such it occupies space ; therefore
(3) There is occupied space ; —

and so with every such argument. Here again generalization produces a catenate syllogism. Moreover, we have already seen (Chap. XXII.) that relational sequences often have complex antecedents, and that in this case they are naturally stated in peculiar syllogisms of their own. For example,

A is greater than B ; and
B is greater than C ; therefore
A is greater than C.

5. Let us now enumerate the leading modes of fallacy; and, after that, discuss them.

(1) In simple translative inference, which is expressed by the "hypothetical" syllogism, two modes of paralogism are to be avoided — that of the *denied antecedent*, and that of the *asserted consequent*.

(2) The disjunctive syllogism, which is a complicated mode of the translative, is also subject to two modes of error — these may be named the *omitted alternative*, and the *false contrary*.

(3) In relational inference mistakes occur either through some *false assumption*, or through some *confusion* respecting a premise or a conclusion.

(4) Fallacies in probable inference arise when *the contingent is confounded with the necessary*, or vice versa; or when *the ratio of the chances is wrongly computed*. In mere possibility there may be *false assumption* of a condition; or of an antecedent as containing one.

(5) Paradigmatization may be erroneous either because of a *misunderstood precedent*, or because of a *false comparison*.

(6) Principiation may show either *premature interpretation*, or credulous *theory-worship*.

(7) The applicative syllogism may have a *false sumption* (major premise), or a *false subsumption* (minor premise).

(8) Finally, catenate syllogizing may be defective either in the *combination of its premises,* or in *one or more of its sequences* (or propositions) *considered singly.*

6. The hypothetical syllogism is interesting, because its peculiar closeness to the fundamental law of all reasoning gives to its rules an universal applicability. Any argument whatever may be thrown into hypothetical form; and then we immediately see that it is composed of a reason and a consequent. Some confine the hypothetical syllogism to apodeictic sequence, but there is no ground for any such limitation.

The first rule of inference is that *if the antecedent be asserted the consequent may be asserted, but that if it be denied, nothing follows.* Therefore to deny the antecedent and to claim a consequence, either positive or negative, is a paralogism. Exact apodeictic sequences, such as are based on definitions and mathematical conversions, form an exception to this rule; in these we either assert or deny either antecedent or consequent, and in every case have a consequence. But this exception does not destroy the rule.

The second rule of inference is that, *in any apodeictic sequence, the denied consequent is followed by the denied antecedent.* But, with the exception mentioned above, to assert the consequent, in order to produce an apodeictic converse, is fallacious. Because "iron is fusible," we cannot say "what is fusible is iron." The Greeks termed this error the ὕστερον πρότερον; since it puts that first which should be last.

But the asserted consequent is not fallacious if we infer only contingency, and say "what is fusible may be iron."

7. The "omitted alternative" and the "false contrary" are the fallacies of disjunctive reasoning: they can be easily understood if we consider, first, the modes of disjunctive assertion, and then the modes of disjunctive inference.

Disjunctive *assertion* arises when an antecedent supports a plurality of alternative consequents; and it may be either strong or weak. In strong disjunction the alternatives are contraries of one another; as when we say, "The season is either spring, summer, autumn, or winter." In weak disjunction the alternatives are compossible; they may exist together, though they cannot all be non-existent at once. Thus, in saying above, "Catenate syllogizing may be defective either in the combination of premises or in one or more of its sequences considered separately," we do not mean that it may not have both these faults at the same time. To say, "The man is either a fool or a knave or a fanatic," means that he has one of these characters at least, not that he has one only.

Disjunctive *inference*, also, has two modes, the positive, or the *tollendo ponens;* and the negative, or the *ponendo tollens* (Chap. XVIII.). The former of these employs for "major" premise either a strong or a weak disjunction, indifferently. On denying all the alternatives but one, we assert that one absolutely; or, on denying several conjointly, we can assert the rest disjunctively. Evidently, in either case, every alternative must be taken into consideration. Our only ground of positive assertion respecting one or several alternatives, is that all the rest have fallen out of the race. Therefore the "omitted alternative" is a fatal defect in any case of *tollendo ponens.*

On the other hand, in the *ponendo tollens*, we assert one contrary alternative, and then deny conjointly as many of the rest as we please; there is no need that we should think of all. But we must now guard against the paralogism of the "false contrary." The *ponendo tollens* is valid only with true contraries. We cannot say that, because the man is a fool, therefore he is not a knave or a fanatic.

Such argumentation, indeed, can be rendered valid if the alternatives be conceived as exclusive of one another, and another alternative be added for the possible union of two or more of the single alternatives. If the man be either simply a knave or a fool or a fanatic, or more than one of these at once, then, if one of these things be true, each of the rest is false. This, however, so alters the reasoning that it is no longer the same argument.

---

8. Relational inferences — even when they have complex antecedents and may be analytically syllogized — are extremely simple, and not in themselves liable to error. Fallacies in

metaphysical and mathematical reasonings arise from *primary misapprehensions*, and from *confusions of thought;* rather than from any difficulty in ontological sequence itself. For instance, the metaphysical doctrine held by some that there are activities without an agent, and powers without a substance, springs partly from the unfounded prepossession that our faculties of immediate cognition perceive only "phenomena," and not also those permanent factors by which actions and changes are produced; and partly from a confusion which takes things to be separable because they can be separately conceived of and mentioned. So also mathematical error originates in concealed confusion and assumption; not through any mistake in axiomatic sequence.

The fraction $\frac{1}{1-x}$ when developed, gives the endless series $1 + x + x^2 + x^3 + \cdots + x^n$. Substituting the value 3 for x in the fraction and in the series, and placing the results in equation, we have $-\frac{1}{2} = 1 + 3 + 9 + 27 + 81 + \cdots$, and so on indefinitely. Thus a quantity less than nothing, or rather a definite subtractive quantity, appears to be equal to an additive quantity indefinitely large.

The paralogism of this lies in failing to note that, no matter to what extent the series may be developed, there will always be a remainder with which the series must be terminated. If we stop dividing when x^3 has been obtained, there will be a remainder requiring $\frac{x^4}{1-x}$ to be added to the quotient, *i.e.* to the series. If we stop with x^4, $\frac{x^5}{1-x}$ must be added. In other words, if we stop with 27, the quantity $-\frac{81}{2}$ will be required to complete the series, or if we stop with 81, the quantity $-\frac{243}{2}$ will be required. But, combining either of these with the sum of the preceding terms, we obtain $-\frac{1}{2} = -\frac{1}{2}$; and all contradiction disappears.

That men do not err in their immediate ontological intuitions, but only through some wrong apprehension of premises, or through some confusion or commutation of ideas, seems to be the doctrine of Aristotle, when he contrasts δόξα, which may be erroneous, with ἐπιστήμη, which is "always true" ("Analyt.

Post.," II. 19). It is also the basis of the paradox of the Latin poet,

"Nam neque decipitur ratio, nec decipit unquam."

9. In problematic sequence, fallacies in possibility do not call for separate consideration. The inference of the possible, when it is *orthological*, may be classed with those relational inferences of which we have been speaking, and, when it is *homological*, is subject to the accidents of homologic syllogizing. But probability, and contingency, which is indeterminate probability, require a specific discussion.

Probable inference has this in common with the disjunctive, that in both we conceive of the contingent and conflictive consequents of one antecedent; in other respects these modes of sequence differ widely. Till comparatively recent times probability has been treated as either not admitting or not needing analysis; and has not been distinguished from contingency. Even now, though understood by philosophers and mathematicians, it has scarcely secured a place in "formal" logic. But it has a distinct nature of its own; and, as we have seen, may be expressed by a peculiar syllogism. In this the first premise sets forth an antecedent which supports a limited number of chances, or possible individual consequents; the second determines a ratio between the chances for some specific consequent and the whole number of chances; and the conclusion asserts that consequent with the corresponding degree of likelihood.

Such being the case, two general forms of error are noticeable. We may either mistake the character of the antecedent, and *suppose that to be necessary which is only contingent;* or we may *miscalculate the ratio of the chances.*

The substitution of the necessary (or the impossible) for the contingent, is the fault of those who either believe too easily that every exigency has been provided for, or who give way to despair while there is yet ground for hope.

This, also, is the defect of that philosophy which makes no distinction between extreme probability and absolute necessity. What is very highly probable is sometimes called "morally certain"; yet the opposite of it is entirely possible, and, under peculiar conditions, may become probable. But the opposite of the necessary is impossible.

Moreover, absolute impossibility does not arise from confliction with any instituted order of things; for that order may be changed or interrupted; but from confliction with the laws of being, or the necessary nature of things. Unprejudiced consideration should enable one to say whether an alleged fact or event be impossible in this way or not.

When an event is seen not to be impossible, its probability cannot be properly determined, unless the circumstances of the case be fully considered. Probability varies wonderfully with variations in the antecedent. Nor is it sufficient to note that a thing has happened often in a certain case, or, for some other reason, has many chances in its favor; we must estimate the *ratio* of its chances in connection with the definitely ascertained antecedent. To this end a power of close, accurate, and dispassionate consideration must be developed; otherwise one's judgments in contingency will be without weight, and little more than plausible conjectures. Moreover, in the mathematical compounding of probabilities, much care is needed if we would avoid intellectual displacements ("The Human Mind," Chap. XXIV.).

10. While Aristotle teaches that every inferential process may be stated syllogistically, *he distinguishes paradigmatic and inductive syllogisms from those in which we reason from general, or universal, statements.* He shows that they express modes of inference quite different from those of the ordinary categorical syllogism. ("Analyt. Prior.," II. 25, 26). According to him, arguments from example (παραδείγματα) are not based on the relation of the universal to the particular (as ordinary syllogisms are), nor on that of the particular to the universal (as inductions are); but on that of the particular, or specific, to the particular, or specific. In short, we reason from one individual, or specific, sequence to another, when the one, but not the other, of these has been already ascertained.

Aristotle, indeed, teaches that one specific sequence is inferable from another because a general law can be discerned in, and obtained from, a given precedent, or from several given precedents; but this is entirely consistent with his doctrine, that the reasoning is from the specific, or the individual, and not from the general. For not the general principle, but only the specific case, is given.

Moreover, while his words favor the view that inference from example operates through (though not from) the general, he probably would allow that the generalization might be dispensed with, provided there be that exact understanding of the prior sequence upon which a true generalization (or principiation) might be made. Locke distinctly teaches the doctrine that a second specific sequence may be directly inferred from a first; provided the antecedent of the second agree with the antecedent of the first in those respects which are perceived to be essential.

Such being the case, two modes of error are possible in paradigmatic syllogizing. Either the prior sequence, which might be called *the major premise, may be misunderstood*, so that either the antecedent or the consequent of it is wrongly conceived; or, if the prior sequence be correctly ascertained, *the minor premise may falsely assert* that the new antecedent is essentially similar to the old. The best plan to avoid error in doubtful cases, is to exact a definite law from the prior sequence by principiation; and then to test this law, and reason from it. The thoughtful mind naturally adopts this method. And this shows how closely the paradigmatic and the applicative syllogisms are related to each other; and why the former has been overlooked and neglected by logicians.

11. The inductive syllogism, yet more than the paradigmatic, is recognized by Aristotle. He says ("Prior. Analyt.," II. 25) that when there is a middle term we syllogize through that (διὰ τοῦ μέσου), but when there is no middle term, through induction (δι' ἐπαγωγῆς). He even opposes "induction" to "syllogism," — that is, to applicative and catenate inference; in both of which we infer through a middle term.

According to Aristotle all first principles (πρῶται ἀρχαί) are obtained by induction. In the last chapter of his "Posterior Analytics," he begins a discussion concerning first principles by speaking of that *power of perception* whereby we obtain the knowledge of individual facts and truths, — δύναμιν σύμφυτον κριτικήν, ἣν καλοῦσιν αἴσθησιν; — and he finishes the discussion by saying that first principles are obtained by induction from this perception; — δῆλον δὴ ὅτι ἡμῖν τὰ πρῶτα ἐπαγωγῇ γνωρίζειν ἀναγκαῖον· καὶ γὰρ καὶ αἴσθησις οὕτω τὸ καθόλου ἐμποιεῖ. In this passage "induction" has the broad sense of "principiation." The doctrine thus taught is very acceptable to those who now call themselves "Perceptionalists."

Further, it is to be allowed that *gathering and collating are not absolutely essential to induction*. Often a natural law has been ascertained by one demonstrative experiment; while necessary, or ontological, sequences, on account of their forceful simplicity, may be principiated instantly from a single illustration. It can only be said that the inductive syllogism ordinarily reaches its conclusion through the analytic comparison of a collection of instances; because cosmological sequences — or the instituted laws of Nature — cannot, as a rule, be exactly determined in any other way.

Therefore, in the first, or major, premise, the different individual antecedents are set forth as having, by reason of some common nature, the same consequent:

> These oaks, beeches, elms, maples, etc., have roots.

The second, or minor, premise states what this common nature is;

> These oaks, beeches, elms, maples, etc., are trees,

whereupon the conclusion attaches the consequent to the nature, or to the whole logical class as having that nature;

> Trees have roots.

This conclusion is certain when the premises justify an universal rule; probable, if they warrant only a rule with exceptions. Probability is not necessarily, or inseparably, connected with induction; and the homological law, on which all principiation rests, is, in itself, ontological and apodeictic.

Induction, like probability, has many specific rules and modes of fallacy, which cannot be considered in a general logic. But two comprehensive forms of paralogism may be mentioned.

The first of these is the *premature interpretation of the individual sequences;* and this may affect either premise. When we say, in the major, that a, b, c, d, e, etc., have each the consequent C, we must be sure that there is a real consequence in each case; and not a mere accidental connection. To assume a sequence simply because one thing may have happened sometimes in conjunction with another is to substitute chance for law. "Simple enumeration" is only the beginning of the inductive process; and even this enumeration is valueless if it be one-sided, ignoring either negative or affirmative instances. How worthless are the cases with which quacks advertise their nostrums, and on which the superstitious base their expectations of good or of evil! Scientific teachings, too, occasionally are affected with hasty or superficial interpretations. Whenever the operation of causes is complex — for example, in the problems of sociology — only wise discrimination can say what are, and what are not, the consequences of given antecedents. In such cases, abandoning immediate principiation, we should separately analyze and ascertain the operation of each factor. In this way we may at last reach a satisfactory conclusion.

The important part of induction lies in exactly determining and understanding the individual sequences of the major premise; after this, the antecedent of the conclusion, to be asserted as predicate of the second premise, and which completes the interpretation, is obtained by defining that common character which belongs to the individual antecedents; and gives to each its efficacy. This also calls for care and deliberation.

Errors of interpretation are often re-inforced by *a strange credulity with which plausible theories are received.* Never content with ignorance, even when knowledge may be difficult or impossible, men form conjectures about all things. This habit is not irrational; nor are conjectures to be despised, if they be not visionary. Yet it has constantly happened in the history of science that doctrines, which have been at best merely conceivable hypotheses, have been advocated and taught as if they were established truths. This mistake has been committed by men of ability, who have been wanting in maturity of judgment, but who wish to rank among "advanced thinkers"; and even sometimes by distinguished observers of Nature who have not, either theoretically or practically, mastered the laws and limitations of inductive reasoning.

12. The applicative syllogism, because of its simplicity, is comparatively free from liability to error; but it may become fallacious either through a *false "sumption,"* or major premise, or through a *false subsumption,* or minor premise. The former results from some prior fallacy, or misconception, or misstatement; while a deceptive subsumption, in a similar way, either falsely ascribes some character to a subject, or ascribes a character which does not make that subject agree with the subject of the sumption. For, in every correct applicative syllogism, the sumption sets forth a general truth; and the subsumption an individual or singular subject, as having the nature of the subject of the sumption; whereupon, in the conclusion, the predicate of the sumption is asserted of the individual subject.

13. We have now considered the fallacies of unconnected, or separate, as distinguished from those of connected, or catenate, inference. These latter remain to be discussed. They are those immediately related to the Aristotelian syllogism; and which, therefore, have chiefly engaged the attention of logicians.

CHAPTER XXVII.

FALLACIES IN CATENATE INFERENCE.

1. Catenational fallacies are of two classes: (*a*) the interior, (*b*) the exterior; commonly called the "formal," and the "material." 2. They may be subdivided into seven classes. 3. The "four-terms," and the "ambiguous middle." 4. The fallacy of "accident" proceeds "*a dicto secundum quid*," either "*ad dictum simpliciter*" or "*ad dictum secundum alterum quid*"; and may be interpreted in either of two ways. 5. The spurious fallacy of accident is merely a case of equivocation. 6. The fallacy of "negative-premises" has two modes: (*a*) "the uncontradicted-middle," and (*b*) "the unasserted-middle." 7. "Illicit-process of the major," relates to negative syllogisms. 8. "Illicit process of the minor," relates to contingent syllogisms. 9. The "undistributed-middle," is (*a*) fatal to the operation of the "separating-consequents," (*b*) weakening to that of the "consequent-consequent." 10. Is to be avoided in the first figure by an apodeictic major; but in the third figure an apodeictic minor suffices. Summary of doctrine respecting this fallacy. 11. The fallacy of two contingent premises is essentially the same with the "undistributed-middle." 12. An apodeictic conclusion cannot follow a contingent premise; hence universal affirmatives are naturally proved in the first figure only.

1. MOST logical writers take the catenate syllogism, and that, too, in its dogmatic form, as the fundamental type of reasoning; hence they discuss fallacies in connection with it, and as deviations from its laws. This course fails to consider things according to their true differences. To understand fallacies we must separate the catenate from the applicative inference, and yet more from the principiative, the translative, and the other modes. Every form of inference has laws of its own, which may be violated.

Paralogisms in catenate syllogizing are of two general classes. Those of the first class arise from a *wrong combination of premises*, and are such as peculiarly affect the catenation of sequences; those of the second class flow from the *falsity of separate sequences,* and are traceable to causes not specially connected with catenate syllogizing.

With reference to their origin, fallacies of the first description might be named *interior catenational fallacies;* and those of the second, *exterior catenational fallacies.* The former result exclusively from the want of a proper catenation between terms, so that logical sequence fails; the latter wrongly assume terms in a premise, or wrongly substitute terms in the conclusion. In each of these last-mentioned cases there may be true catenation: but in the one the conclusion rests on unwarranted assumption; and in the other the conclusion supported by the premises, is falsely identified with that which ought to have been proved.

These interior and exterior fallacies ordinarily receive the designations "formal" and "material"; the one being supposed to relate exclusively to the nature of logical sequence, and the other to the nature of specific antecedents and consequents. This distinction is inaccurate: both modes of paralogism are directly related to the nature of sequence; and in either mode any particular fallacy must be explained with reference to the matter considered. The one, indeed, brings logical connection into prominence, and the other, the character of the things reasoned about; but neither is exclusively "formal," or exclusively "material."

2. Interior catenational fallacies may be treated under the five following heads:

(1) The fallacy of four terms; and of the ambiguous term.

(2) The fallacy of negative premises; and of an affirmative conclusion with either premise negative.

(3) The illicit process of the major premise; and the illicit process of the minor.

(4) The undistributed-middle: first, as in syllogisms of the consequent-consequent; and secondly, as in syllogisms of the separating-consequents.

(5) The fallacy of a guarded conclusion with both premises particular; and of an apodeictic conclusion with either premise particular.

Exterior catenational fallacies may be treated under two general heads:

(1) The *petitio principii;* in which either premise is assumed without warrant, and

(2) The *ignoratio elenchi;* in which the true point at issue is ignored, and an irrelevant conclusion proved instead.

3. Adopting the order of the above divisions, we begin with the fallacy of four terms (*quattuor termini*). More exactly speaking, this is the fallacy of using more than three terms; it is a violation of the rule that *catenation requires three terms and admits three only.* Other modes of inference employ four terms or more. Should we say,

A is equal to B;
B is equal to C; therefore
A is equal to C,

we would use four terms, A, B, equal to B, and equal to C. This, however, is not a catenation of sequences, but a single sequence which has for subject, or antecedent, "A, being equal to B, which is equal to C," and for predicate, or consequent, "equal to C." So, also, paradigmatic and inductive inferences use more than three terms; and may use even many, if each example, or instance, be counted as having an antecedent and consequent of its own.

On the contrary, catenation with more than three terms is so evidently fallacious that it is never attempted except under some carelessness and confusion. In saying,

A wicked man desires happiness;
The only road to happiness is virtue; therefore
A wicked man pursues virtue,

five terms are used. Yet the argument might be taken as a syllogism of three terms.

In most cases the fallacy of four terms is concealed under some ambiguity, one of three terms being employed in two senses, and serving really as two terms. This occurs with the middle term more frequently than with either of the others; and hence this form of the fallacy is commonly known as the "ambiguous middle." But sometimes, instead of the middle term having different meanings in the premises, the major, or the minor, term, has one meaning in its premise and another in the conclusion.

Fallacies of ambiguity have received various names according to their origin. When the paralogism arises because the same word or phrase has two significations, and is therefore "equivocal," it is called *the fallacy of equivocation*. Thus one might argue that since "criminal actions" should be punished, prosecutions for theft should be punished; because they are "criminal actions." Again, the paralogism based on an ambiguous grammatical construction, is styled *the sophism of amphibology;* for a sentence containing such a construction is termed amphibolous. Thus,

> Twice two and three is seven;
> But ten is twice two and three; therefore
> Ten is seven.

Closely related to such fallacies are those of Division and of Composition. These arise from using an universal expression collectively in one premise, and distributively in the other. In the following we proceed from division to composition, and commit the *fallacy of composition,*

> All the angles of the triangle are less than two right angles;
> *A*, *B* and *C* are the angles of the triangle; therefore
> They are (collectively) less than two right angles.

On the other hand it would be a *fallacy of division* to say,

> All the angles of the triangle are equal to two right angles;
> *A* is an angle of the triangle; therefore
> It is equal to two right angles.

The *fallacy of "accent"* arises when an improper antithesis is attached mentally to some part of a sentence, and expressed by emphasis. "Love the *brethren*" might mean "the brethren, but not strangers"; and so one could say, "He is not my brother; I need not love him."

The *fallacy of the figure of speech* proceeds from the figurative signification of a term to the literal; or from the literal, to the figurative. If our Saviour's words, "This is my body," be figurative, the argument from it in favor of the "real presence" in the Eucharist is illogical.

4. Lastly, among errors of ambiguity, we name the "*fallacy of accident*"; of which there are two forms, the *genuine* and the *spurious*. The former arises when we reason from a nature with an accidental addition, as if it were a nature viewed simply, or *per se* — that is, from what is true specifically, as if it were true universally. In the fallacy of accident we proceed "*a dicto secundum quid*" either "*ad dictum simpliciter*" or "*ad dictum secundum alterum quid.*" For this mode of paralogism assumes that what is true of a species is necessarily true of the genus, or of another species. Should we say,

> What destroys health should not be used;
> Intoxicants destroy health; therefore
> No intoxicants should be used,

this would be a case of "*a dicto secundum quid ad dictum simpliciter*"; because the minor premise does not mean that intoxicants, as such and always, ruin health. But should we say,

> To take life is not sinful;
> Murder is the taking of life; therefore
> Murder is not sinful,

the "taking of life" would be spoken "*secundum quid*" in the major premise, and "*secundum alterum quid*" in the minor. Only certain modes of killing are not sinful; and only certain other modes are murder.

According to the foregoing explanation, the fallacy "*a dicto secundum quid*" (the genuine fallacy of accident), is an error of ambiguity; but it may also be interpreted as an error in distribution, or modality. The thought of the argument being only partially expressed, we may make the fallacious premise either an universal assertion regarding an unnamed and understood species, or a contingent assertion regarding the genus. To say, "*the taking of life with just cause is not sinful*" is specific and apodeictic; to say "*the taking of life is not necessarily sinful*" is generic and contingent. Completing our thought in the former way, without expressing the completion, the fallacy is an ambiguous middle; completing it in the other way, it is a case of "undistributed middle"; a fallacy of which we shall speak presently.

The easier mode of refuting the error seems to be to refer it to an ambiguity. The sophism,

> What grows on sheep is raw wool;
> Those who wear woollens wear what grows on sheep; therefore
> Those who wear woollens wear raw wool,

is best refuted by showing the double meaning of the middle term "what grows on sheep." In one premise this signifies "what grows on sheep in its primitive condition"; but in the other, "what grows on sheep in its manufactured condition." We might, however, adopt the modal interpretation, and deny that "what grows on sheep is (always) raw wool"; in which case the fallacy would be one of distribution.

5. The spurious fallacy of accident does not admit of a double treatment. It is founded on that metonymy whereby the same term indicates — not a genus and a species — but *the whole, and a part, of a nature*. Should we say, in Darapti,

> Man is mortal;
> Man is a rational spirit; therefore
> A rational spirit is mortal,

the term "man," in the major premise, signifies the whole composite being, but in the minor, only a part of that being. This paralogism is said to proceed "*a dicto simpliciter*" ("man" in the major) *ad dictum secundum quid* ("man" in the minor). But erroneously; for we do not proceed from the genus to the species, but from the composite whole to its component part. To proceed from the generic to the specific is not fallacious.

Then, should we invert the order of the premises, and infer

> What is mortal is a rational spirit;

this also would be an ambiguous middle, not a true fallacy of accident.

In addition to the paralogisms just mentioned, certain others have been improperly identified with those of accident. They are reasonings concerning individuals, or singulars, as such; and therefore cannot be said to proceed from the specific to the generic or to the specific. Yet they are closely related to the genuine fallacies of accident, because they use *terms with unexpressed additions*. In the relational syllogism,

> The meat we eat to-day was bought yesterday;
> But that bought yesterday was raw meat; therefore
> The meat we are eating is raw meat,

the middle term, "that bought yesterday," is ambiguous by reason of two additions. But since these are not the accidents of a general essence, the fallacy is not one in catenate syllogizing. It is an erroneous inference in identity.

6. The second general mode of catenational paralogism is that connected with negative premises; and it has two forms. For *to draw any conclusion from two negative premises, and to infer an affirmative conclusion if either premise be negative*, are direct violations of those principles which govern all catenate syllogizing. The law of the consequent-consequent requires the minor premise to be affirmative; and the law of the separating-consequents, that one premise, no matter which, be affirmative; neither admit a conclusion if both premises be negative.

Then, also, the separating-consequents, which has always one premise negative, has always a negative conclusion; while the consequent-consequent has always a negative conclusion if the major premise be negative. Therefore, with either law, a negative premise necessitates a negative conclusion.

In a case of the separating-consequents, the fallacy of two negative premises might be specifically known as "*the uncontradicted-middle*"; because, in this mode of reasoning, one premise must assert, and the other deny, the same consequent term. But with the consequent-consequent, the fallacy takes the form of "*the unasserted-middle*"; because, in this mode of syllogizing, the middle term, which is assumed as antecedent of the major premise, must be asserted as consequent of the minor.

Here, however, we must remember that, with the consequent-consequent, if the antecedent of the major premise be negative, the minor premise must be negative; and can be affirmative only in the sense that it asserts, as its own consequent, the antecedent of the major. In such a case a negative minor premise may be followed by an affirmative conclusion; and also two negative premises, by a negative conclusion. We can say,

> One who is not helpless should be hopeful;
> He who is sound in mind and body is not helpless; therefore
> He should be hopeful.

And also,

> One who is not helpless should not be despondent;
> He who has mental and bodily health is not helpless; therefore
> He should not be despondent.

7. We now come to those modes of paralogism, in each of which a term is improperly used with a distributive, or necessitant, force. These are *the illicit process of the major*, the *illicit process of the minor*, and the *undistributed middle*.

In the first of these the major term is distributed in the conclusion, when it has not been distributed in its premise. Such a process is illegitimate; because it puts more into the conclusion than the premises warrant. A weaker conclusion than can be maintained is not unlawful; but a stronger conclusion is.

Affirmative sequence, in general, — and therefore an affirmative conclusion — does not call for a distributed predicate. Hence, the major term need not be distributed in the premise of an affirmative syllogism. But negative sequence — and therefore a negative conclusion — does distribute its predicate (Chap. XX.); consequently the major term must be distributed in the premise of every negative syllogism. To say,

> Motion is visible;
> Sound is not visible; therefore
> Sound is not motion;

is entirely inconclusive unless we mean that *all* motion is visible. Hence, in negative syllogisms of the second and of the fourth figure, the major premise must be universal; for, in these figures, the major term is subject of that premise.

But in negative syllogisms of the first and of the third figure, the major term is predicate of the premise. In such syllogisms, therefore, the major premise need not be universal in order to avoid the illicit process; but it must be negative; for only negative assertion distributes the predicate. Hence, the mood Bokardo is lawful; and such conjectural syllogisms in the first and third figures, as the following:

> Some hard things are not metals;
> { Some hard things are minerals (3d fig.), or
> { Some minerals are hard things (1st fig.); therefore
> A mineral may not be a metal.

8. The illicit process of the minor distributes the minor term in the conclusion, when it has not been distributed in the premise. This paralogism is less violent than the illicit process of the major. The latter syllogizes when no inference at all is warranted; illicit process of the minor concludes apodeictically when only a contingent conclusion can be justified.

This fallacy will scarcely deceive thoughtful persons, unless it should be in the fourth figure. But should we say,

> No Hindoos are negroes;
> All negroes are blacks; therefore
> No blacks are Hindoos,

there would be an illicit process of the minor. The term "black" has greater modal force in the conclusion than it has in the premise. The proper conclusion would be,

> Some blacks are not Hindoos.

9. The "undistributed-middle" occurs when the middle term is undistributed in both premises; but not if it be distributed in either. While common to all modes of catenate inference, it does not affect all alike; but is a fallacy which has two degrees, a stronger and a weaker.

In the first place, a distributed-middle is necessary to *any conclusion whatever by the separating-consequents*. This principle requires the middle term to be predicate of both premises; and one of the premises to be negative. Therefore, the middle term must be distributed, as predicate of the negative premise; hence, a distributed-middle is indispensable in every negative syllogism of the second figure.

This is true, also, of negative syllogisms in the fourth figure. For these immediately fall into the second figure, on the conversion of the minor premise; and may be interpreted as really governed by the separating-consequents. To say, in Fresison,

> No moral principle is an animal impulse;
> Some animal impulses are principles of action; therefore
> Some principles of action are not moral principles,

is essentially an argument in Festino; and would be wholly inconclusive without the distribution of "animal impulse."

The relation of the undistributed-middle to negative syllogisms in the fourth figure, may be stated, also, from a more radical point of view. In the ultimate analysis these syllogisms involve the mental conversion of both premises; one of which must be negative. But this involves a distributed-middle. For a negative premise, to be convertible, must be universal and distribute both terms, one of these being the middle term.

What distinguishes negative syllogisms in the second and fourth figures from all others, is that they depend on the principle of conversional contradiction; which is, "contradict an apodeictic consequent, and you may deny the antecedent." In the second figure the minor premise contradicts the consequent of the major; and thereupon the conclusion denies the antecedent of the major. In the fourth this same process takes place, if we mentally convert the minor premise only; but if we convert both premises, and appeal directly to that principle in which all syllogizing originates, we must then use the "denied-consequent" in converting the negative premise, whichever that may be. In either case conversional contradiction is involved; and in either this requires a distributed-middle. For if we reduce to the second figure, then, according to the law of that figure, the middle term will be distributed as predicate of the negative premise; while reduction to the first figure by converting both premises is conditioned on the universal negative premise; in which the middle term is distributed either as subject or as predicate. This absolute need of a distributed middle affects only those syllogisms which depend on conversional contradiction.

The second, and weaker, form of the "undistributed middle" pertains to syllogisms of the first and of the third figure, and to the affirmative moods of the fourth figure — in short, to all syllogisms which are governed by the consequent-consequent alone or with some simple conversional addition. In such reasonings the undistributed-middle merely renders the conclusion unguarded, or conjectural. It is a paralogism only when employed to prove a guarded conclusion; or any conclusion that can be expressed dogmatically.

Distribution of the middle does not of itself ensure a guarded conclusion; in any syllogism whatever, if either premise be unguarded, the conclusion will be unguarded. But this second mode of the undistributed-middle assumes that both premises are guarded; and is the attempt to reach a guarded conclusion from two guarded premises, without a distribution of the middle term. Moreover, as there is a sense in which the stronger mode of the error belongs to reasonings by the separating-consequents, so there is a sense in which this weaker mode is confined to syllogisms of the consequent-consequent. For all syllogizing may be said to take place on one or other of these two principles.

10. To avoid this fallacy in the first figure, an apodeictic major is necessary. The middle term, as predicate of an affirmative proposition, not being distributed in the minor premise, it must be distributed, if at all, as subject of the major. To say,

>Some virtues are laudable;
>Some habits are virtues; therefore
>A habit may be laudable,

yields a correct unguarded conclusion; but we must say,

>All virtues are laudable;
>Some habits are virtues; therefore
>Some habits are laudable,

if we would have a guarded conclusion.

The reason for this is that if A (habit) be followed in any mode by B (virtue), and B necessarily by C (laudable), then A is followed in the same mode by C as by B. On the contrary, if A be followed in some mode by B, but B only contingently by C, it may turn out, notwithstanding a justifiable conjecture, that A never is, and never can be, followed by C. In other words, without distribution of B the conclusion cannot be guarded against impossibility.

Here, however, it is to be remembered that a guarded conclusion can be reached in a peculiar way, if the middle term be distributed in the minor premise, even while that term may be undistributed in the major. We can say,

> Some heroes are godlike;
> Some men are all the heroes; therefore
> Some men are godlike.

For if C (godlike) in some way follow B (hero), and B necessitate A (man), then C must follow A as it does B. But, *ex hypothesi*, C follows B with a guarded contingency; therefore the conclusion is guarded.

This form of reasoning, nevertheless, though logically conclusive, is properly excluded from the first figure; because we do not naturally use a contingent antecedent with a necessitating consequent, but always put the necessitant first; even when some mental conversion must follow, in our syllogistic use of the assertion. We do not say, "Some men are all the heroes," but, "All heroes are men." Hence we reason in the third figure, instead of the first, and say,

> Some heroes are godlike;
> All heroes are men; therefore
> Some men are godlike.

Clearly in this case the third figure is really a modification of the first; and the chief function of its affirmative minor premise is to distribute the middle term, when it has not been distributed in the major. And this is equally true respecting the positive moods of the fourth figure; each of which can distribute the middle term only as the subject of the minor premise.

. The law respecting the distribution of the middle term may be summed up as follows: All catenate syllogisms are divisible into two classes; those which depend on conversional contradiction, and are essentially syllogisms of the separating-consequents; and those in which that principle is not employed, and which are essentially syllogisms of the consequent-consequent. In the former the distributed middle is absolutely indispensable; in the latter it is necessary only for guarded conclusions.

Yet, in thus opposing syllogisms of the separating-consequents to those of the consequent-consequent, we really contrast one operation of the consequent-consequent with another; the separating-consequents being equivalent to the consequent-consequent with the addition of a conversional contradiction. Our teaching, here, is not inconsistent with the doctrine that the consequent-consequent is the fundamental principle of catenate inference. It refers all syllogizing to this law as having, or as not having, that conversional addition.

11. The fallacy which may result from *two particular* (or contingent) *premises*, is radically the same with that of the undistributed-middle. In syllogisms of the separating-consequents no conclusion whatever can follow two such premises; because, in the second figure the major premise, and in the fourth the negative premise, must be apodeictic. Precisely the same reasons which demand a distributed middle require one premise at least to be universal. But in syllogisms of the consequent-consequent, if both premises be particular, a conjectural conclusion is lawful; only a guarded conclusion is fallacious. This is the weaker form of the paralogism.

Here, however, a particular affirmative, used as minor premise in the first figure, if its predicate be distributed, must be regarded as an universal affirmative, and as really relegating the argument to the third figure. For, in saying,

> Some heroes are godlike;
> Some men are *all the heroes;* therefore
> Some men are godlike,

the minor premise means "all heroes are men," is apodeictic in effect, and justifies the guarded conclusion. This shows that the rule requiring one premise to be universal, in order to a guarded conclusion, signifies, speaking exactly, that the middle term must be distributed in one or other of the premises.

12. The last fallacy to be named, as arising from false syllogistic construction, is the inference of *an apodeictic conclusion if either premise be contingent.* The self-evident truth, that a chain of connection cannot be stronger than its weakest part, applies to every mode of syllogizing. Accordingly, if we use the separating-consequents, which is conditioned on an universal major premise, the minor must also be universal; and if we use the consequent-consequent, both premises, likewise, must be universal; if the conclusion is to be universal.

Indeed, normally, *universal affirmative* conclusions not only require two apodeictic premises, but also that these be so related to each other, that the antecedent of the minor may become the antecedent of the conclusion, and the consequent of the major the consequent of the conclusion. Hence, that form of proposition, which is the most important of all, the universal affirmative, is provable properly only in the first mood of the first figure.

Such, then, are the fallacies in syllogistic connection, whether arising from too many terms, or ambiguous terms, or negative premises, or undistributed terms, or particular premises.

CHAPTER XXVIII.

EXTERIOR CATENATIONAL FALLACIES.

1. Exterior catenational fallacies are of two modes, (*a*) the *petitio principii* and (*b*) the *ignoratio elenchi*. 2. The direct *petitio* is called the *non-causa*, and also "the false, or fictitious, middle." 3. Specific forms of it are the *non-tale pro tali*; and the *post-hoc, ergo propter hoc*. 4. The indirect *petitio* includes (*a*) the *implicatio mendax*, (*b*) the *circulus in probando*, and (*c*) the *saltus in deducendo*. 5. The *ignoratio* may take place either with or without intention. Often accompanies a shifting of of the ground. The *elenchi mutatio*. The *argumenta ad hominem, ad populum, ad verecundiam*. 6. Fallacies "in dictione" and "extra dictionem,"—a superficial distinction.

1. As already stated, there are modes of paralogism by which a conclusion can be wrongly supported even while there may be a correct catenation of sequences. Though one's syllogism be perfect, his argument will be worthless, if either his premises be unreliable, or if he prove something different from that which he ought to prove, and which is the point at issue.

The first of these sophisms is known as the *petitio principii*, or "begging of the question"; the second, as the *ignoratio elenchi*, or the irrelevant conclusion. In both of the Latin names there is reference to a question as under debate; this indicating that these fallacies occur more frequently in discussions of long standing, than in new investigations.

The phrase "petitio principii" means the illicit assumption of some principle, or ground of inference; and intimates that this mode of paralogism, though it begs the question, does not immediately assert the point at issue. An immediate assertion could scarcely be fallacious, because it would not have even the appearance of argument.

2. The *petitio* may be defined as the paralogism of false assumption. This, however, does not signify that the premise may not be true, but only that it is unwarranted.

No premise can be fallaciously asserted without some show of reason; and this appearance may either appeal formally to our faculties of knowledge, or may be supported by concealed implications. There are, accordingly, two general modes of the *petitio*, the direct and the indirect.

The former of these is the more common and important; and is known by the name "*non-causa pro causa.*" In this expression the word "*causa*" must be taken, in a broad, logical sense, to denote an antecedent, or ground of inference; whether it be an efficient cause or not. Logic does not consider efficient causes as such. The "*non-causa*" is a syllogism, in either premise of which something is falsely assumed to be the antecedent of a given consequent.

This paralogism is also called, more technically, "*the false, or fictitious, middle*"; and is thus contrasted with the ambiguous, and with the undistributed, middle. For, from the nature of the case, if either premise, or both, be falsely assumed, the middle term must be falsely used either as subject or as predicate or as both.

There is also another reason for this second designation. The middle, or connecting, term, is, in a pre-eminent sense, the cause, or ground, of the conclusion; τὸ μέν αἴτιον τὸ μέσον, says Aristotle ("Analyt. Post.," II. 2). Hence, if the middle, either as antecedent or as consequent, be fictitious, there is really nothing — no reason — to produce conviction.

Should we say, "The magnet is animated, because it moves itself," there would be a *non-causa*, or fictitious middle, on the supposition that the implied major, "Whatever moves itself has life," is not true, or not evident. And, even were this allowed to be true, there would still be a fictitious middle unless there were reason to believe that "magnets move themselves."

3. Specific forms of the *non-causa* have received specific names. When the false premise is supported by a superficial resemblance, but has no true analogy, to some known sequence, it is styled the "*non-tale pro tali*"; as, for example, that "a bat must lay eggs; because it has wings, and flies." Also, if, without any true reasoning about causes, a general law is asserted simply because one event has, more or less frequently,

preceded another, the fallacy is designated the "*post hoc, ergo propter hoc*"; as that "protection (or free-trade) must be a good policy, because countries have prospered under it." They may have prospered despite of it.

4. We now pass to those forms of the *petitio*, which employ the aid of indirection and concealment. They are three in number; and may be named the fallacies of the false implication, of reasoning in a circle, and of the gap in argument; or, using Latin terms, the *implicatio mendax*, the *circulus in probando*, and the *saltus in deducendo*.

The unfair implication is the device of those who do not immediately assert a falsehood, but tell things, or raise questions, which presuppose the falsehood, as if it were true. When the words of such persons are made plausible by reason of an ingenious accommodation to facts, this method of deceit often proves successful. It is a favorite instrument of political warfare; and appears in those lying inventions which demagogues circulate, to deceive the public concerning the character of statesmen and the designs of parties.

That mode of the *implicatio mendax* which logicians notice most, is rather amusing than important. It is the fallacy of the second, or implicating, question (*sophisma heterozeteseos — fallacia plurium interrogationum*). An ancient example of it, named the "cornutus," employs the query, "Have you cast your horns?" This is really a second question, which implies that the prior enquiry, "Have you had horns?" can be answered affirmatively. Hence, if one replies that he has not cast his horns, it will be said, "Then you have them yet!" while, if he says that he has cast them, it can be said, "Then you were once a horned animal!"

A specific mode of the heterozetesis, is the *fallacia plurium interrogationum*, in which several disconnected questions are asked together; as if all could be answered at once and in the same way. In saying, "Are honey and gall sweet?" it is presupposed that these substances have the same taste, whatever that may be. This presupposition must be rejected as unwarranted; after that, the enquiry can be answered as two questions.

All paralogisms of the heterozetesis are refuted by showing that a false implication, attached to the nature of the question, renders a categorical answer absurd and illogical.

The *circulus in probando* arises when a premise is at first reasoned from hypothetically, and then afterwards proved by using the conclusion itself as a premise. Thus a speaker might argue that a policy is wise, simply because it works well, without giving sufficient evidence regarding its working; and afterwards that it must work well, because it is wise. The defect of such a procedure is commonly concealed by length of argument and new forms of expression.

The "*saltus*," or leap, in ratiocination, occurs when some sequence, in a series, does not really follow upon the preceding one, yet is assumed to do so. Such an argument employs a succession of middle terms, each of which is consequent to a foregoing, and antecedent to a following, term. The confessed absence of one of these connections would rob the reasoning of all logical force; evidently, therefore, the *saltus*, when analytically stated, must participate in the nature of the fictitious middle, or *non-causa*. Indeed, speaking broadly, this latter paralogism would include every mode of the *petitio*. But the *saltus* is distinguished specifically from the *non-causa* proper, because the fallacy of it is hidden in the midst of a succession of inferences, most, or all, of which may be correct.

5. The last general mode of fallacy to be discussed is the *ignoratio elenchi*, or irrelevant conclusion. This takes place when somehow the question at issue is misstated, and a conclusion proved different from that really required.

Often this paralogism is preceded or accompanied by a "*shifting of the ground*" of the argument; either with or without, intention. For, although the same proposition may be maintained on different grounds, a change in reasons sometimes indicates that the conclusion first attempted is being abandoned for another.

When the *ignoratio* happens inadvertently, it is called "missing the point"; but it is frequently a piece of sophistry. In either case there is an aptness in the word "ignoratio"; for this suggests a mental activity in the rejection of knowledge, which is more than simple "ignorantia," or ignorance.

"Elenchus" (ἔλεγχος) originally signified the proposition to be maintained in refutation of an adversary — the contradictory of his assertion — and then came to mean, in general, the conclusion to be proved, the question at issue.

The *ignoratio* takes place, not only when the substituted conclusion is entirely new, but also when part only of an assertion is proved, as if it were equivalent to the whole; for example, that a man was influenced by money, and (being so) was mercenary; or that he killed another, and (in doing so) committed murder; or that some measure is open to certain objections, and should be rejected (no proof being given that the disadvantages outweigh the advantages).

The point in dispute is sometimes altered at the beginning of a discussion; more frequently this change takes place during the course of debate; in which case we may use the specific name, "*elenchi mutatio*," or a change of the question. This substitution may occur through mere confusion of thought in abstruse discussions; but it is chiefly to be found in the reasonings of those who wish to have some controversy decided apart from its true merits. Thus the *argumentum ad hominem* seeks to confound an adversary by showing his inconsistency, selfishness, or want of principle, instead of proving the unworthiness of his cause. Such reasoning is opposed to the *argumentum ad rem*, or *ad judicium*. It is never admissible in judicial proceedings; and can be permitted in political discussions only when the public interests require the exposure of incompetent and unreliable leadership.

The *argumentum ad populum* and the *argumentum ad verecundiam* are also, for the most part, mere sophistries. The former of these is a demagogic appeal to vanity and ignorant prejudices, in a case where the requirements of justice and right should be presented; the latter excludes arguments based on truth, by urging the respect due to persons of reputation or authority. So, also, the *argumentum ad ignorantiam* is a demand that some opinion shall be accepted, because one's adversary has nothing better, or more plausible, to offer. All these forms of the *mutatio* are mentioned by Mr. Locke, in the fourth book of his essay (Chap. XVII.).

6. In the chapters which are now brought to a close, we have not referred to a distinction made by Aristotle, and adopted by many logicians as a basis for their discussions concerning fallacies. It signalizes the fact that the deceptive power of some paralogisms depends partly, or wholly, on forms of verbal expression; while that of others is independent of this, but arises exclusively from the character of the thought employed. Aristotle, accordingly, distinguished fallacies *in dictione* from fallacies *extra dictionem*.

This division is useful as reminding us that frequently, in order to refute a paralogism, it is necessary to define the meaning of words and the force of constructions, while, in other cases, this work is not needed, but only a determination of the thought. At the same time, the distinction is too superficial to form a basis for thorough exposition. For, after language has been explained, fallacies *in dictione* are found to resolve themselves into fallacies *extra dictionem*.

INDEX AND VOCABULARY.

This index gives the name of every author mentioned in the foregoing discussions, and the number of every page on which he is mentioned.

It is also designed to assist the student, who may be interested in any particular point, or question, to trace the teachings of the book respecting that point, as these may present themselves in the successive chapters. In other words, it is offered as a kind of concordance.

In addition, the intention has been to include every technical logical term, with references to the pages on which its meaning is explained; and, in this way, to furnish a defining vocabulary.

For further information respecting metaphysical terms, or topics, especially as related to the *perceptionalist* philosophy, the reader is referred to the author's "Human Mind" and "Mental Science."

ABSTRACT, the term, 46.
Abstraction, distinguished from generalization, 37.
Accent, fallacy of, 307.
Accident, as a "predicable," separable and inseparable, 58.
Accident, as opposed to substance, 35, 58.
Accident, fallacy of, genuine and spurious, 308.
Accidents of entity, 121.
Accidental definitions, 63.
Action (ποίειν) as a category of predication, 50.
Actualistic belief and assertion, 24, 87, 104, 150.
Adjunct, defined, 59.
Affirmation, 81, 90, 93.
Affirmative, the, does not always affirm, 248-254, 310.
Affirmative propositions, their conversion, 192-3.
αἴσθησις, or perception, the basis of all knowledge, 300.
"All," collectively and distributively, 43, 307.
Ambiguous-middle, the, 306.
Analogy, of natural sequences, 142; inference from, 172.
Analytic judgments, 125.
Animals, divided logically, 75, 76.
Antecedent and consequent, the law of, 7, 9, 107, 111, 123, 131, 151, 152-3.

Antecedent, exact and reciprocative, 111.
Apodeictic, or demonstrative, inference, 23, 108, 150.
Applicative inference, 131, 140, 152, 226, 303.
Apprehension, simple, 33.
Argumentum, ad rem vel judicium, ad hominem, ad populum, ad verecundiam, ad ignorantiam, 322.
Aristotle, his organon, 2; his modal propositions and syllogisms, 97, 101, 241; his definition of judgment and the proposition, 6, 82; on contingency, 8; his definition of the syllogism, 10, 222; his categories of predication, 46; his use of the term, οὐσία, 48; on species and definition, 54; on genus and difference, 55, his distinction between the internal and the external word, 84; his use of the term "categorical," 85; interprets "not" to signify separation, 90; defines necessity, 110; on the origin of knowledge, 119, 300; on the law of contradiction, 126 ; on paradigmatic inference, 132; on final causes, 144; his four causes, 144; discusses only three syllogistic figures, 236 ; prescribes no order of syllogistic statement, 240; his dictum, 263; his reduction of syllogisms, 282-6; on example and induction, 299; on first

325

principles, 300; on the middle term, 300, 319; his division of fallacies, 323; *vide* "Mental Science."
Art, the term as applied to logic, 13, 16, 26.
Ascript, ascripta, defined, 36, 48, 26.
Ascriptional predications, 50, 92, 268.
Asserted-consequent, the, 194, 200, 243, 252.
Assertion, defined, 80, 153.
Assertivity, 93.
Assertory, a term used by Kant, 98.
Attribute, defined, 59.
Axioms, 121, 133; syllogistic, 262-7.

"BEGGING the question," 318.
Being, or entity, 51, 58.
Belief, or conviction, 21, 79.
Binomial formula, the, in the calculation of probabilities, 168.
Bowen, Prof. Francis, on modality, 4, 101.

CANONICS, the Epicurean name for logic, 174.
Canons of experimental enquiry, 145.
Categorematic and syncategorematic words, 34.
Categorical propositions, 85, 88, 89, 106, 174.
Categories of Aristotle, the ten, 46.
Category, defined, 46.
Catenate inference, 227-8, 304; fallacies in, 303-320.
Cause and effect, the law of, 139, 146.
Certainty, moral, 172.
Chances, defined, 9, 162; the ratio of the, 164; calculation of, 164-9.
Cicero, the inventor of the word "essence," 67.
Circulus in probando, the, 320.
Class notion, the, 43.
Clearness, defined, 61.
Common-antecedent, the, 232, 251, 265-6.
Common-consequent, the, 250.
Common-sense, doctrine of, 5.
Comparison, 37.
Composition, the fallacy of, 307.
Compound assertions, 92.
Conception, 31.
Conceptualism, 38.
Concrete, the term, 46.

"Conditional" assertions and reasonings, 81, 85, 89, 103, 106, 156.
Conditionative, or modal, propositions, 97, 99, 102, 106, 174.
Conditions, doctrine of, 8, 104, 109, 110, 111, 131.
Conditions, necessitant, or logical, 111.
Confliction; see contrariety.
Conjugation, or syzygy, of syllogisms, 240.
Consequent-consequent, the, 229, 259, 263, 279.
Consequent, the asserted, 194, 200, 243, 252.
Contingency, 8, 9, 24, 101, 117, 180, 183, 197, 204, 292, 298.
Contingency, in the wide sense and as including possibility, 101, 183, 189.
Contingency, half-guarded, 182, 186, 197, 208, 218, 246, 264; unstable, or unguarded, 184, 209, 219, 248, 258, 273; guarded, 185, 314; encouraging and discouraging, 187; empirical and mathematical, 198, 207, 210, 264; fixed, or embedded, 220.
Contingent syllogisms, 244, 248, 289.
Contradiction, the law of, 121, 126, 128.
Contradiction, consequential and categorical, 156, 157.
Contradictory opposition, 177, 182, 187; in a wide sense includes contrariety, 155.
Contradictories, are conceivable only in pairs, 158.
Contraposition, explained, 127, 193, 203.
Contrariety, 154, 157, 176, 180, 296.
Conversion, logical, 190; of necessary sequences, 111; ground for, 125; of particular negatives, 193.
Conversion *per accidens*, or by the asserted consequent, 194, 200, 243, 280, 283; *per differentiam*, or by the retained-necessitant, or differential conversion, 194, 200, 244, 280, 315; "simple," 194; by negational exclusion, 194; by the denied-consequent, 200, 243, 280, 313; of contingency, 201, 211, 215; as related to the law of reason and consequent, 243.
Conviction, 22, 79.
Co-ordination, logical, 76.
Copula, origin and use of the verb "to be," as 89; *vide* "Mental Science."

INDEX AND VOCABULARY. 327

Cosmological judgments; *vide* "Mental Science."
Creational development, a possible theory, 143.
Criticism, scientific, 140.

DEDUCTION, 133.
Definite and indefinite notions, 42.
Definitions, 54, 61, 67; distribute the predicate, 96.
De Morgan, Prof., his definition of logic, 14.
Demonstration, 23, 108, 150.
Denial, 153.
Denied-consequent, law of, 157, 200, 243, 282.
Design in nature, 144.
Dichotomy, as a mode of division, 78.
Dictum of Aristotle, the, 262.
Difference, individual or numerical, 40, 56; specific, 40; as a predicable, 55; *vide* "Mental Science."
Dilemma, the, constructive and destructive, 160.
Discourse of reason, 18.
Disjunction, logical, 77, 154, 159, 256, 296.
Disjunctive syllogism, 159, 296.
Distinctness, as a quality of thought, 61.
Distribution, the, of a notion, 43, 93.
Division, the fallacy of, 307.
Division, logical, not didactive nor rhetorical, 30, 72; a synthetic process, 71; rules of, 72-77; expressed by a predication, 96.
Dogmatic, the term, 98, 242, 261.

Elenchi mutatio, 322.
Elenchus, defined, 322.
Embedded contingency and possibility, 183, 220.
Enthymeme explained, 275.
Entity defined, 31, 51.
Enumeration, simple, 302.
Enunciation, as distinguished from assertion, 80.
Epicheirema, the, 275.
ἐπιστήμη, always true, according to Aristotle, 297.
Episyllogism, 278.
Error, origin of, 290.

Essence, 55, 67; singular, 69; the nominal and the real, 69.
Essential and accidental definitions, 62.
Euler's symbolic diagrams, 268.
Excluded-middle, the law of the, 121, 128.
Exclusive and exceptive assertions, 97.
Exercises in constructing and reducing syllogisms, 287.
Existence and non-existence, 31, 153.
Experience, or ἐμπειρία, as including every immediate perception, 300; for other meanings of the term see the author's "Mental Science," Chap. XLIX.

FACT is both positive and negative, 32.
Factual propositions, 103, 261; their conversion, 190.
Fallacies, 290; formal and material, 305; in *dictione* and *extra dictionem*, 323.
False, or fictitious, middle, 319.
Falsity, or untruth, an ambiguous expression, 291.
Figures of syllogism, the, 238, 248, 262, 279.
Figure, the fourth, 238, 254, 262, 267, 281; as compared with the others, 259, 282, 311-314.
Final cause in nature, 144.
Form and matter, 68; *vide* "Mental Science."
"Formal" logic, 28.
Formal or schematic notions, 33.
Fundamentum, the term, 74, 75.

GASES, logically divided, 75.
Generalization, defined, 37; the principiative, 132.
General notion, 37.
Genus, as a predicable, 53, 55, 155, 159; the predication of it not necessarily analytic, and either individual or general, 53.
Goclenius, Rudolphus, professor in Marburg, his form of the sorites, 277.
God, causally unconditioned, yet logically necessary, 110; *vide* "The Human Mind," Chap. XXI.
Guarded and unguarded contingencies, 182, 186, 197, 208, 218, 246, 248, 249, 253, 255, 258, 281-3.

HAMILTON, Sir William, his definition of logic, 3; his quantification of the predicate, 95; on modal propositions and syllogisms, 262; his syllogistic notation, 270; his doctrine of the syllogism, 274.
Historical, or factual, propositions, 103.
Homologic inference, 7, 18, 130, 150, 225.
Hypothesis, the inductive, 139.
Hypothetical, or suppositive, judgments and propositions, 24, 87, 103; vide "Mental Science."
Hypothetical syllogism, the, 150, 151, 156, 228, 294.

IDEA, now used as equivalent to notion, or conception, 31.
Idealism, defined, 39.
Identity, the law of, 121, 124, 128, 190.
Identity, numerical and specific, 41.
Ignoratio elenchi, 318.
Illation, or inference, and illative, or inferential, 7, 98, 103.
Immediate, or intuitive, knowledge, 138, 290; vide "Mental Science."
Implicatio mendax, the, 320.
Impossibility, 99, 100, 113, 186, 298.
Indefinite notions, 42
Indefinite quantity, 94.
Individuals and individual notions, 40-44.
Induction, 132; the act, 136; the process, 137; canons of, 145; the inductive syllogism, 300.
Inference, and inferential propositions, 6,-103, 105, 150; immediate, 108, 119, 293.
Inherential propositions, 82.
Intuition, 18; vide "Mental Science."
Irrelevant conclusion, 321.

JUDGMENT discussed, 6, 79.

KANT, Immanuel, his views on logic, 1; his definition of judgment, 6; his term "assertory," 98.
Knowledge, or absolute and well-founded conviction, a species of judgment, 79; vide "Mental Science."

LAMBERT, an excellent German logician, 270.
Leibnitz, Gottfried Wilhelm, on the category of substance, 47.

Linnaeus, his definition of man, 63.
Locke, John, his definition of judgment, 6, 22, 83; of reason, 17; on the category of substance, 47; on the nominal, and the real, essence, 69; on substance, 70 ; on the origin of knowledge, 119; on immediate inference, 119; on certain fallacies, 322.

MAJOR, minor, and middle terms, 238.
Mark, the added, as the basis for logical division, 75.
Mathematical principles, 122; inference, 297.
Matter, in logic, 68.
Maxims of inductive conjecture, 141.
"May" and "may not," 182.
McCosh, James, Pres., quoted, 21.
Mediate and immediate inference, 108; as a distinction in relational inference, 224.
Mental and verbal propositions, 84, 86, 95, 103.
Metaphysical first principles, 122.
Metaphysics, or ontology, the basis of logic, 4, 120.
Methodology, defined, 27.
Methods, of agreement, of difference, etc., 146-9.
Mill, John Stuart, on inductive methods, 146; on contingency and probability, 207.
Mnemonic lines of Petrus Hispanus, the, 283.
Modal, or conditionative, predications, 97; as contracted with the pure or dogmatic, 102, 106 ; essentially illative, 174.
Modalist, reason for this name, 4.
Modality, 10, 101, 106, 150, 242, 262.
Modus ponendo ponens and tollendo tollens, 153; tollendo ponens and ponendo tollens, 159, 294.
Moods, syllogistic, 11, 240, 242, 262, 287; their symbolic notation, 270.
Mortgage investments, 77.

NATURE of a thing, the, 55, 59.
Nature, or the universe, the intellectuality of, 141-4.
Necessity, 99, 180; logical, 101, 113, 186; its converse, 199; vide "Mental Science."

Necessitant, the retained, 194, 244, 247, 253; condition, the, 111.
Negation, 81, 90, 93.
Negative propositions, 94; must sometimes be construed as affirmatives, 245, 252, 254, 310.
"No," the adjective, 94.
Nominal and real definitions, 66.
Nominalism, 39.
Non causa pro causa, the, 319.
Non-existence and the non-existent, 32.
Non tale pro tali, 319.
"Not," the particle, 91, 95.
Notational definition, the, 65.
Notion, the, an idea, or conception, named from its relation to knowledge, 31.

OBJECT, objectivity, objectuality, 31.
Ontology, as related to logic, 4, 120, 301.
Opposition, logical, 174, 179.
οὐσία, substance, or essence, 46.

PARADIGMATIC inference, 131, 299.
Paralogism, defined, 290.
Parsimony, the "law" of, 143.
Particular propositions, 269; categorical and modal, 93, 175, 196, 197.
Perceptionalism, defined; *vide* "Mental Science."
Perceptions, simple and immediate, free from error, 290.
"*Per se*" and "*per accidens*" explained, 285.
Petitio principii, the fallacy of, 316.
Place, or position, as a category of predication, 49.
Plurium interrogationum, the fallacy, 320.
Polysyllogism, defined, 278.
Ponendo ponens and *ponendo tollens*, 153–9.
Positing, or assertion, of a statement, the, 153.
Position, or posture, as a category of predication, 50.
Possession, or condition, as a category of predication, 50.
Possibility, 113, 173, 205, 293; as including contingency, 99, 115; embedded, 114, 183; unstable, 184.
Possible to be, the, 113; and the possible not to be, 115.

Post hoc, ergo propter hoc, the fallacy of, 317.
Postulates, 121, 133.
Predicables, the five, defined, 52; their use, 59.
Predicate, defined, 35, 85, 89, 95.
Predication, 80, 83; grammatical, distinguished from logical, 84; force of categorical, 95.
Predicative notions, 34, 45.
Premises defined, 237; order of, 240; false conclusion from true, etc., 291.
Presentational perceptions, 23.
Presentential propositions, 82, 87.
Principiative inference, or principiation, 131, 226, 300.
Principle, the term, 73; principles of inference, 119, 133.
Principium individuationis, 41.
Probability, 99, 116; conditioned on possibility, 100; its oppositions, 189; ordinary and philosophical, 170; orthologic and homologic, 171; the calculation of, 164–9; *vide* "Human Mind," Chap. XXIV.
Problematic inference, 23, 108, 150, 292, 298.
Property, as a predicable, generic, and specific, 57.
Propositions, 79–88, 104.
Proprietal conceptions, 58.
Prosyllogism, 278.
Pure, the term, 28, 98, 242.
Pure, or dogmatic, propositions, 97, 98, 242, 261; verbal in character, 102; as related to modal, 106; pure, or dogmatic, syllogisms, 261.

QUALITY, as a category of predication, 48.
Quality, as a predicable, 60.
Quality of propositions, the, 81, 93, 175.
Quantity, as a category of predication, 48.
Quantity of propositions, the, 93; a kind of added predication, 102, 175.
Quantification of the predicate, 92, 274.
Quantification of modals, the, 195.

RATIO of the chances, the, 100, 116, 162, 198, 298.
Ratiocination, or reasoning, 108.
Real definitions, 66.

Realism and Nominalism, 39, 41; vide "Mental Science."
Reason, the faculty of, 16; the intuitive and the discursive, 17.
Reason and consequent, the law of, 7, 9, 10, 107, 131, 152, 260.
Reasoning, 108; in the general, 134; inductive, 136.
Reciprocating necessities, 95, 111.
Reciprocation, the law of syllogistic, 233, 267.
Reductio ad absurdum, 127, 287.
Reduction of syllogisms, 278; new method of, 282; indirect reduction, 283; *per impossibile*, 286.
Reid, Dr. Thomas, on modal syllogisms, 262.
Relation, as a category of predication, 49; vide "Mental Science."
Relations, logical, or necessary, 110.
Relational, or adjunctional, definitions, 63.
Relational inferences and syllogisms, 223, 293, 296.
Representative essence, the, 68.
Retained-necessitant, the, 194, 244, 281, 315.

Saltus in deducendo, 320.
Schematic notions, 33.
Scholastic definition, the, 65.
Separate, and catenate, inference, 290, 304.
Separating-consequents, the law of the, 230, 249, 260, 265, 279.
Shifting the ground of argument, when fallacious, 319.
Simple and compound assertions, 92.
Simplicity of Nature, the, 144.
Singular notions, 42, 69.
Singular propositions, really have no "quantity," 94.
Solidity, primary perception of, 139.
Sophisma heterozeteseos, the, 320.
Sophistry, involves the intention to deceive, 290.
Sorites, the, 275; the Goclenian, 277.
Species, the predicable, sets forth the whole nature conceived of, 54; as a class-name, 155.
Specific difference, 40; may, like genus, be either individual or general, 53.
Sphere of logic, the, 3, 25.
Square, the logical, 176.

Stoic doctrine of final cause, the, 144.
Subcontrariety, 178, 188.
Subordination, or subalternation, of propositions, 125, 176, 177, 181, 184.
Subject, the term, 20, 35, 85, 89.
Subjective and predicative notions, 34.
Subjectual, the term, 20.
Sublation, defined, 153.
Substance, as a category of predication, 45; vide "Mental Science."
Substance and accident, as corresponding to subject and predicate, 35, 58.
Substance, *metaphysical*; vide "Mental Science."
Substanta and ascripta, 36, 45, 51, 69.
Substantal predications, 50, 92, 125, 268.
Substantal and ascriptional predicates, 60.
Substantialization of ascripts, 51, 191, 268.
"Substantial form," 69.
Substitutional judgments, based on the law of identity, 124.
Sumption and subsumption, 294, 303.
Superalternation; see subalternation.
Supposition, or hypothesis, 139.
Suppositive, or hypothetical inference, distinguished from the hypothetical syllogism, 151.
Syllogism, the Aristotelian, 7, 10, 135.
Syllogism defined, 222, 292; the syllogism proper, 7, 223, 228, 237, 274; the disjunctive, 159, 294-5; the translative, or hypothetical, 151, 228, 294; the relational, 223, 296; the paradigmatic, 225, 299; the principiative, 226, 299; the inductive, 136, 300; the applicative, 226, 303.
Symbolization of contingencies, 220.
Syncategorematic words, 34.
Synthetic judgments follow the principle of identity, 124.
Synthetic order of premises, the, 240, 278.

TERMS, or extremes, 237; middle, major, and minor, 238.
"The," as article, sometimes has a distinctive, without a singularizing, force, 38.
Theophrastus, the immediate successor of Aristotle: his use of the term "categorical," 85.

Thought, or conception, and belief, or conviction, the primary powers of mind, 21; *vide* "Mental Science."
Tollendo tollens, 153; and *tollendo ponens*, 159.
Transfer, the principle of logical, 151, 154, 294.
Truth, defined, 18; hypothetical, 150.
Tychologic principle, the, or ratio of the chances, 162.

UNASSERTED-MIDDLE, the fallacy of the, 309.
Uncontradicted-middle, the fallacy of the, 309.
Undistributed-middle, the, 311.
Uniformity of natural operations, 142.
Unital, the term, 41, 94.
Unity, or oneness, defined, 40.
Universals, impossible entities, 39; *vide* "Mental Science."
Universal propositions, 93; their conversion, 192; modals, 195, 196.

VERBAL and mental propositions, 84, 86, 95, 103.

WOOLSEY, Pres. Theodore D., his divisions of international law, 72.